SEX, REPRODUCTION AND DARWINISM

To Alister my friend and accomplice, who helped make this book possible

Filomena

Cambridge, 10 December 2014

SEX, REPRODUCTION AND DARWINISM

EDITED BY

Filomena de Sousa and Gonzalo Munévar

PICKERING & CHATTO
2012

Published by Pickering & Chatto (Publishers) Limited
21 Bloomsbury Way, London WC1A 2TH

2252 Ridge Road, Brookfield, Vermont 05036-9704, USA

www.pickeringchatto.com

BRITISH LIBRARY CATALOGUING IN PUBLICATION DATA

Sex, reproduction and Darwinism.
1. Reproduction. 2. Sex. 3. Evolution (Biology)
I. Sousa, Filomena de. II. Munevar, Gonzalo.
571.8-dc23

ISBN-13: 9781848932647
e: 9781848932654

This publication is printed on acid-free paper that conforms to the American
National Standard for the Permanence of Paper for Printed Library Materials.

Typeset by Pickering & Chatto (Publishers) Limited
Printed and bound in the United Kingdom by the MPG Books Group

CONTENTS

ACKNOWLEDGEMENTS

The editors wish to express their thanks to Pickering & Chatto for its interest, first suggested by Daire Carr, for the guidance provided by Ruth Ireland and the support from Eleanor Hooker. The volume was made possible by the authors who accepted with enthusiasm our invitation to contribute their original work, and who put much time and effort into writing and revising their chapters. We also wish to thank Yi Zheng for her dedicated editorial assistance, as well as Clifton Johnson for his great help with many of the figures. Filomena de Sousa is indebted to the Foundation for Science and Technology of Portugal (FCT) for financial support and owes great many thanks to the University of Cambridge for its immense generosity in granting access to varied resources and providing the intellectual means that allowed her to pursue her work. Gonzalo Munévar is indebted to his wife, Susan, for her extraordinary patience and her loving encouragement during the long process of editing the volume.

LIST OF CONTRIBUTORS

Pieter R. Adriaens is postdoctoral fellow in the Institute of Philosophy at the University of Leuven where he received his PhD. He held visiting fellowships at the Institut des Sciences de l'Evolution in Montpellier, and at the University of Cambridge in the Department of History and Philosophy of Science, and the Centre for Research in the Arts, Social Sciences & Humanities (CRASSH). His main interests are in history and philosophy of psychiatry, sexology and biology. With Andreas De Block, he edited a book on philosophy, psychiatry and evolutionary theory (*Maladapting Minds* (2011)), and a special issue on the history of evolutionary psychiatry (*A Hundred Years of Evolutionary Psychiatry, 1872–1972*; *History of Psychiatry* (2010)).

Jens Bast is a PhD candidate in the group of Professor Scheu at the Georg-August-University Goettingen, working in evolutionary ecological genomics. His work focuses on transposable elements in sexual and putatively long-term asexual oribatid mites, spermatogenesis and a comparative study of mitochondrial genomes of mites. These projects incorporate new DNA-sequencing technologies and bioinformatic analyses into evolutionary investigations. Previous work involved screening for parthenogenesis-inducing Wolbachia in oribatid mites at Darmstadt University.

Julia Sandra Bernal is Professor of Law at Universidad del Norte. She received her PhD in Law and Bioethics from UNED (Spain). She has taught law at Universidad del Norte since 1994, primarily civil law and bioethics. She has extensively worked on evolutionary theory, human rights issues and bioethics. Select publications: 'Clonación: Un fundamento evolucionista de los principios de dignidad e igualdad frente a la técnica de clonación de seres humanos con fines reproductivos', *Revista de Derecho*, 26 (2006) Universidad del Norte; 'Nuestra naturaleza como especie biológica: la razón de una posibilidad y una necesidad para una propuesta ética común', *Revista Prisma Jurídico*, 9 (2010); and *Moral y derecho en la evolución humana* (2011).

William M. Brown is Senior Lecturer of Biomechanics and Psychology at the University of Bedfordshire. He received his PhD from Dalhousie University, followed by a NSERC Postdoctoral Fellowship at Long Island and Rutgers Universities. His research focuses on developmental stability, genomic imprinting, epigenetics, cooperation and conflict. Select publications include: 'The Parental Antagonism Theory of Language Evolution: Preliminary Evidence for the Proposal', *Human Biology*, 83 (2011), pp. 213–45; W. M. Brown, L. Cronk, K. Grochow, A. Jacobson, K. Liu, Z. Popovic and R. Trivers, 'Dance Reveals Symmetry Especially in Young Men', *Nature*, 438:7071 (2005), pp. 1148–50; W. M. Brown, M. E. Price, J. Kang, N. Pound, Y. Zhao, and H. Yu, 'Fluctuating Asymmetry and Preferences for Sex-Typical Bodily Characteristics', *Proceedings of the National Academy of Sciences of the United States of America*, 105 (2008), pp. 12938–43.

Lucrecia Burges is Associate Professor at the University of Balearic Islands and Visiting Scholar at the University of California, Irvine. Her research includes the evolution of the human mind, moral evolution and sexual dimorphism. She is member of the Laboratory of Human Systematics and the research group on Human Evolution and Cognition, and participated in ten research projects on the evolution of cognitive processes. Select publications include: 'Evolutionary Epistemology a Clue to Understand Moral Origins', *History and Philosophy of the Life Sciences*, 24:1 (2002), pp. 109–20; 'Natural Values or Taking Contributions to Morals Seriously', *History and Philosophy of the Life Sciences*, 24:2 (2002), pp. 275–84; and C. Cela Conde, L. Burges, M. Nadal and A. Olivera, 'Altruisme et impartialité: selection non-naturelle?', *Comptes Rendus Biologies*, 333:2 (2010), pp. 174–80.

Camilo Cela-Conde is Full Professor of Anthropology at the University of Balearic Islands and received his PhD in philosophy from Barcelona University. His research covers human systematics (C. Cela-Conde and Ayala, 'Genera of the Human Lineage', *PNAS*, 100 (2003), pp. 7684–9, and brain correlates of higher cognitive functions ('Sex-Related Similarities and Differences in the Neural Correlates of Beauty', *PNAS*, 106 (2009), pp. 3847–52. He is a fellow of the American Association for the Advancement of Science, Salk Institute and University of California, San Diego. Select publications include: *On Genes, Gods and Tyrants* (1989); and *Human Evolution* (2007). He has led several research expeditions including: Tugen Hills, Baringo Lake District, 2005; and Natron Lake, 2006, research on Hadza hunter-gatherers.

Andreas De Block is Assistant Professor of Philosophy at the University of Leuven, where he received his PhD. He studied philosophy and psychology in Leuven and Ghent, and was post-doctoral researcher at Radboud University

Nijmegen and the University of Michigan. He is the author and editor of several books on psychoanalysis, philosophy of psychiatry, and philosophy of biology. His main research interests are cultural evolution, the philosophical assumptions of the nature-nurture debate, sexual orientation and philosophy of psychiatry.

Eve-Marie Engels is Full Professor of Ethics in the Life Sciences in the Departments of Biology and Philosophy at the University of Tübingen where she heads the DFG-Research Training Group 'Bioethics – On the Self-Design of Human Beings by Use of Biotechniques' at the International Centre for Ethics in the Sciences and Humanities. She received her PhD in philosophy from the Ruhr-Universität Bochum, and was professor of philosophy at the University of Kassel, guest professor at several universities, and was awarded a Heisenberg-grant. Her research includes ethics, theory and history of biology. She has authored more than 150 articles, three monographs, including *Charles Darwin* (2007), and is editor/co-editor of fourteen books, including with Thomas F. Glick, *The Reception of Charles Darwin in Europe*, 2 vols (2008). She served on the German National Ethics Council (2001–7) and the Ethics Committee of HUGO International.

Jagdish Hattiangadi is Professor of Philosophy and Natural Science (Science and Technology Studies) at York University, Toronto. He received his PhD from Princeton University in history and philosophy of science, earned his MA with distinction from the London School of Economics in 1965, and BA (Hons) from Bombay University in 1962. He is currently writing a book (*The Theory and Craft of Breaking Through in Science*), in an effort to rehabilitate an unjustly discredited inductive method. He has published widely in diverse areas in philosophy, in the history of science and on problem solving in the sciences. His work is partially listed at www.yorku.ca/jagdish. Since 1971, he has been the editor of the journal *Philosophy of the Social Sciences*.

Victor S. Johnston is Emeritus Professor of Psychology at New Mexico State University. He received his PhD in psychology from the University of Edinburgh (where he was granted the A. E. Bennett Neuropsychiatric Research Foundation Award), followed by postdoctoral fellowships at Yale and Stanford University. His major area of research is human emotions examined from behavioural, evolutionary and electrophysiological perspectives. His work on the biological basis of human facial beauty received worldwide recognition in the form of television shows in the United States, Europe and South America, international publications, and a book *Why We Feel; The Science of Human Emotions* has been translated into multiple languages. His computer software for evolving human facial images (FacePrints) awarded a US patent, 'The Inventor of the Year Award', from the New Mexico Entrepreneurs Association is now used by the police throughout Great Britain. His major publications are listed in wikipedia.org.

Ken Kraaijeveld is a post-doctoral researcher at the Leiden Genome Technology Center, Leiden University Medical Center. He has worked at research institutes in California, Melbourne and London. His current research focuses on the consequences of Wolbachia-induced parthenogenesis in parasitoid wasps and is applying state-of-the-art genome sequencing technology to study how asexuality affects the control of transposons in their genome. Past projects included studies on a wide variety of organisms: birds, flies and marsupials. His work has been published in a wide variety of scientific journals, such as *Evolution and Molecular Ecology*. He is associate editor of the ornithological journal *Ardea*.

Elisabeth A. Lloyd is Maxine and Arnold Tanis Chair of History and Philosophy of Science, as well as a Professor of Biology, and holds Adjunct positions in Philosophy and the Center for the Integrated Study of Animal Behavior, as well as being an Affiliated Faculty Scholar at the Kinsey Institute and Associated Faculty of the Cognitive Science Program. She received her PhD from Princeton supervised by Bas van Fraassen, having also studied with Richard Lewontin (Harvard, Genetics, 1983–4). She worked at the University of California, San Diego until 1988, and UC Berkeley until 1998, when she joined Indiana University. Her books, *The Structure and Confirmation of Evolutionary Theory* (1998/1994, finalist for Lakatos Award), *Keywords in Evolutionary Biology* (co-edited with Evelyn Fox Keller, 1992), *The Case of the Female Orgasm: Bias in the Science of Evolution* (2005, Bullough Award) and *Science, Evolution, and Politics* (2008), as well as her forthcoming *Philosophical and Conceptual Issues in Climate Modeling* (co-edited with Eric Winsberg), reflect her foundational interests.

Her early research interests have focused around models and confirmation in evolutionary biology, and her book *The Structure and Confirmation of Evolutionary Theory* (1994; 1998), was a finalist for the Lakatos Award. She has also edited *Keywords in Evolutionary Biology* with Evelyn Fox Keller (1992). She pursued her theoretical ideas about natural selection with Stephen Jay Gould and published two articles with him. Her interests in philosophy of biology, general philosophy of science and gender issues are pursued in her book, *The Case of the Female Orgasm: Bias in the Science of Evolution* (2005), and she has been given the Bonnie and Vern Bullough Award from the Foundation for the Scientific Study of Sexuality. Her writings have been published in a variety of philosophical and scientific journals and edited books. Some of her essays are collected in her recent book, *Science, Evolution, and Politics* (2008). She is currently co-editing a new book, *Philosophical and Conceptual Issues in Climate Modeling*, with Eric Winsberg, for Chicago University Press. As well as holding the Maxine and

Arnold Tanis Chair of History and Philosophy of Science, she is a professor of biology, and holds adjunct positions in philosophy and the Center for the Integrated Study of Animal Behavior, as well as being an Affiliated Faculty Scholar at the Kinsey Institute and Associated Faculty of the Cognitive Science Program.

Gonzalo Munévar is Professor of Humanities and Social Sciences, Lawrence Technological University. He received his PhD in Philosophy of Science at UC Berkeley under the supervision of Paul Feyerabend. As a visitor he has done research or taught at many universities around the world, including Stanford, the University of Washington, Irvine, Barcelona, Edinburgh, Newcastle (Australia) and Kobe Shodai (Japan). He is presently engaged in experimental and theoretical research in cognitive neuroscience. Other interests include evolution, space exploration, philosophy of science and writing fiction. He has published many books, including *Radical Knowledge* (1981); *Evolution and the Naked Truth* (1998); *The Worst Enemy of Science?* (ed. with J. Preston and D. Lamb, 2000); and the novel *The Master of Fate* (2000). A list of publications is available at http://www.ltu.edu/arts_sciences/humanities_ss_comm/Dr_Gonzalo_Munevar.asp.

Marcos Nadal is Associate Professor at the University of Balearic Islands where he received his PhD in psychology, followed by a José Castillejo post-doctoral grant at the University of Vienna. His research focuses on integrating psychological, neuroscientific and evolutionary perspectives on human derived traits, including aesthetic appreciation, language and morality. He is author of over fifty articles and book chapters, listed on http://evocog.org/en/members/89-marcos.html. He serves on the editorial board of the journal *Psychology of Aesthetics, Creativity and the Arts.*

Lesley Newson is Research Associate at University of California, Davis and Research Fellow at the University of Exeter where she received her PhD. She previously had a two-decade career as a science writer and broadcaster. Her desire to pursue a career in academic research and writing was inspired by the puzzle of why people in modern societies have small families. Her interests have since expanded to the broader implications of new understandings about the cooperative nature of human parenting and the development and testing of cultural evolutionary hypotheses to explain the rapid cultural change that accompanies economic development.

David Reznick is Professor of Biology at the University of California, Riverside, and received his PhD from the University of Pennsylvania. His research focuses on the experimental study of evolution in natural populations and quantification of the genetic basis underlying their differences, the evolution of placentas and Poeciliidae reproduction. He has authored over 125 articles and book chapters, including: *The Origin Then and Now: An Interpretive Guide to the Origin* *of Species* (2010); D. N. Reznick, M. Bryant, D. A. Roff, G. Ghalambor and D. E. Ghalambor, 'Effects of Extrinsic Mortality on the Evolution of Senescence in Guppies', *Nature*, 431 (2004), pp. 1095–9; D. N. Reznick and R. E. Ricklefs, 'Darwin's Bridge between Microevolution and Macroevolution', *Nature*, 457 (2009), pp. 837–42. He is a fellow of the American Association for the Advancement of Science and the American Academy of Arts and Sciences, was vice-president of the American Society of Naturalists (2011). He has also served on the editorial boards of *Evolution*, *Ecology*, *American Naturalist*, *Functional Ecology*, and earned the E. O. Wilson Naturalist Award (2003).

Filomena de Sousa is Research Associate in the Centre for Philosophy of Science at the Faculty of Sciences, University of Lisbon. She received her PhD in philosophy from Université du Québec à Montréal followed by FCT post-doctoral and research fellowships at ISEG- Technical University of Lisbon, and is Visiting Scholar at the University of Cambridge. Her research includes evolutionary and cognitive concepts in social sciences and economics, ethical implications of biological and social engineering in human reproduction, eugenics, as well as topics in history and philosophy of science and method in social sciences and economics as reflected in publications and lectures listed on http://cfcul.fc.ul.pt/equipa/3_cfcul_elegiveis/filomena_sousa/filomenasousa.htm.

Ronald de Sousa is Professor Emeritus of Philosophy at the University of Toronto, and a fellow of the Royal Society of Canada. He received his PhD from the University of Princeton and is the author of *The Rationality of Emotion* (1987); *Évolution et Rationalité* (2004); *Why Think? Evolution and the Rational Mind* (2007) and *Emotional Truth* (2011), as well as of over 100 articles and book chapters. He has lectured in some twenty countries. His current and recent research interests are reflected in a sample of some six dozen articles and reviews available at http://www.chass.utoronto.ca/~sousa. These focus on emotions, evolutionary theory, aesthetics, ethics, cognitive science and sex. He is currently working on a *Very Short Introduction to Love* for Oxford University Press.

LIST OF FIGURES AND TABLES

INTRODUCTION

Filomena de Sousa and Gonzalo Munévar

The purpose of *Sex, Reproduction and Darwinism* is to bring together scholars from a variety of disciplines in biology, the humanities and the social sciences to take a fresh look at some of the main Darwinian themes concerning sex and reproduction. Sexuality and reproduction are, of course, central to evolutionary thinking and, in turn, have come to be conceptualized differently because of developments in evolutionary thinking.

In recent years, and particularly on occasion of the 150th anniversary of the publication of Charles Darwin's *On the Origin of Species* (1859) in 2009, the scholarly and the educated public have been treated to the publication of many books on evolution, including the complete collection of Darwin's works by Pickering & Chatto, and the publication of several international conference proceedings in honour of Darwin's thought. Darwin's thought and the scientific revolution it has wrought, however, spans over such a wide intellectual spectrum that no single volume can possibly attempt to do it justice. Our book thus focuses on two basic concepts in Darwin's approach to biology: sex and reproduction. Although by contrast with, say, the recent volumes on conference proceedings, the focus of our book is narrower, this is by no means a disadvantage. Experimental discoveries and theoretical investigations of sex and reproduction, some made by the authors of this book, have significant implications for our understanding of evolution and for establishing the relationship between evolution and other areas of human endeavour. By concentrating on those discoveries and investigations, we trust we are able to offer a new and challenging, at times controversial, collection of related insights about Darwin's heritage.

Those insights are to be found in the interplay of the several themes of the book, which include the tension between survival and sex and reproduction; the advantages and disadvantages of sexual reproduction over asexual reproduction; how the evolution of sexuality has led it to acquire functions other than reproduction; how sexuality and reproduction have affected the evolution and development of the brain; how sexuality and reproduction have shaped the evolution of human morality and culture; and how ideas and misinterpretations of the evolution of sex and reproduction have affected human society.

These overlapping themes have been organized into the following six sections, with the chapters in each section having been chosen to offer different points of views, sometimes contrary as in the section on sexual selection and the evolution of morality, sometimes complementary as in the section on eugenics.

Darwin was a great experimentalist, and thus it is fitting that we begin Part I: Reproduction, Mortality and Evolution with the work of an experimental evolutionary biologist, David N. Reznick. Reznick finds guppies from the Caribbean Island of Trinidad useful for studying the process of evolution by natural selection from an experimental perspective and for testing evolutionary theory in natural populations. Reznick's experiments are performed both in his laboratory and in the guppies' natural environments. Guppies are found in high and low predation environments which differ in the species of predators that guppies co-occur with. This contrast is found repeatedly in different drainages and the different predation regimes are often right next to one another, separated by a waterfall. Guppies from high predation environments experience much higher mortality rates. High mortality is associated with earlier maturity, a higher rate of investment of resources in reproduction, and the production of more and smaller offspring. All of these differences have a genetic basis. Mortality rates can be manipulated by either introducing guppies from high predation localities into sites from which they and their predators had previously been excluded by waterfalls, thus lowering mortality rates, or by introducing predators into low predation sites over barrier waterfalls, thus increasing mortality rates. Such experiments have shown that life histories evolve as predicted by theory and in a fashion that is consistent with earlier comparative studies. They have also shown that evolution by natural selection can be extremely fast – on the order of four to seven orders of magnitude faster than inferred from the fossil record! This is a truly remarkable result from Reznick's experimental work.

Chapter 2, by Ronald de Sousa, provides an interesting theoretical account of the connection between sex and mortality. Among the most significant achievements of philosophy in the past half-century is the solution to the problem of understanding biological function and teleology in a way that allows it to have objective reality, rather than being necessarily interest-relative. Such is the virtue of the aetiological account of teleology. This holds, roughly, that an organ's effect counts as a function providing that effects of that sort enhanced fitness in the ancestors of that organ, in such a way as to provide a partial explanation for the existence of the present organ and its capacity to produce the effect. However, what happens when this view confronts the claim that our death serves a purpose? It has sometimes been claimed that the biological function of death is to enable evolution, by making room for future generations of increasing complexity and sophistication. Literally understood, de Sousa claims, that view is naïve and absurd. The null hypothesis is that we die simply because greater longevity of individuals is not worth the extra work required, from the point of view of the genes transmitted by any given organism. It is the interlocking of four features of sexually reproducing species that ensure that individual organisms must die not long after reproducing these four features are: metazoan organization, sexual reproduction, division of labour between sex and somatic cells, and individual death.

In Part II: Reproduction without Sex, Ken Kraaijeveld and Jens Bast bring their experimental work to bear in Chapter 3, in which they point out that the evolution of sex remains one of the major enigmas in evolutionary biology. Asexual reproduction occurs in many different organisms scattered throughout the tree of life, yet asexual lineages rarely persist over long periods of time. The answer to why sex prevails in the vast majority of multicellular organisms is often sought in the genetic consequences of asexuality. The lack of recombination is expected to reduce the efficacy of natural selection in removing deleterious mutations, transposon insertions and other genomic reorganizations. However, it is important to realize that asexuality is achieved in a wide variety of ways, ranging from self-fertilization to ameiotic budding. These mechanisms have a profound impact on the genetics of a population, each in their own way. For example, some modes of asexuality fix the genome in a heterozygous state, while other asexuals are completely homozygous. Furthermore, the modifications that are necessary to achieve asexuality may have side effects that affect the fitness of the organism. Last, extraordinarily efficient DNA-repair mechanisms appear to allow some organisms to survive as asexuals for millions of years. Kraaijeveld and Bast discuss the effects of different types of asexual reproduction on genome architecture and genome processes and discuss how these might contribute to the demise of most asexual lineages.

In Chapter 4, Jagdish Hattiangadi examines the role of bacteria in making possible the existence of eukaryotes. As he reminds us, Darwin's theory of evolution by natural selection fails to account for the origin of the sheer abundance and variety of species while assuming blended inheritance. The New Synthesis, interpreting Mendelian genetics within Darwin's theory of evolution, gave us the tools to understand the abundance and variety found in the forms of life. While the New Synthesis is true, as far as it goes, it would seem to be superficial, 'maya' or 'illusion'. Hattiangadi encourages us to look at two sets of troubling facts related to (1) the twofold cost of sex (Maynard Smith) and (2) the symbiogenetic origins of eukaryote cells (Lynn Margulis.) These facts taken together suggest that eukaryotes are a manifestation of the radiation of a particular kind of prokaryote, the purple sulphur bacterium. It evolved into the mitochondrion, an organelle within the eukaryote cell. The twofold cost of sex is not a double cost to the mitochondrial DNA, only to the nuclear genome. Rules governing prokaryote genetics may be murkier, compared to Mendelian laws, but when all is said and done, the natural selection of prokaryotes remains the true underlying story of evolution on earth. The manifest variety is illusory.

Part III: Sex without Reproduction? begins with Chapter 5 by Pieter R. Adriaens, Andreas De Block and Lesley Newson, who argue that modern male homosexuality is a social construction of the last two hundred years, and that three evolutionary approaches provide the best way to elucidate this historical transition, while avoiding the essentialist trap of holding that such sexual orientation is an immutable and discrete trait with clearly defined boundaries. The first approach argues that male homosexuality is (part of) an adaptive strategy that is only triggered in certain circumstances, thus explaining the remarkable historical and cross-cultural prevalence patterns of same-sex sexual behaviour. A second approach builds on the ideas and insights that have been developed in recent (Darwinian) theories of cultural evolution and gene-culture co-evolution. These theories might help us to explain how biologically evolved tendencies and preferences, including male homosexuality, are influenced by cultural learning. A third and final approach attempts to offer an evolutionary explanation for why both scientists and lay people, including many homosexuals themselves, tend to take an essentialist attitude towards human sexual orientation. Although these three evolutionary approaches to constructivism about sexual orientation can – in principle – be reconciled with each other, they are not equally cogent.

In Chapter 6, Gonzalo Munévar reminds us that some important counterexamples to Darwin's *On the Origin of Species* were offered by animal behaviour that decreases either the individual's chances for survival or for reproduction. Altruism is an example of the first, homosexuality of the second. Given the apparent success of explaining altruism using kin selection, E. O. Wilson and others attempted similar explanations of homosexuality as an adaptation. Munévar argues, however, that biological explanation of the existence of homosexuality need not depend on the action of natural selection. Darwinian variation would have most phenotypic traits distributed in a population, and thus sexual preference may be so distributed. Moreover, genes that interact with the environment and with noise, as the genes relevant to sexual preference are likely to do, yield a continuous distribution curve in the expression of the trait. According to LeVay and others, several stages of development have to work in a precise manner for the individual organism to have a clear preference for the opposite sex. This is seen, for example, in the importance of receiving the right levels of testosterone at the right times to differentiate brain structures as male, placement within a litter in a rat's placenta, and the presence of appropriate play environments in young monkeys. Since both the environment and noise may interfere in the development of any one particular stage, we should expect a certain amount of deviation from the mean in a population. Considerations about gene expression, thus, lead Munévar to conclude that homosexuality may be a biological phenomenon without having been selected for.

In Chapter 7, Elisabeth Lloyd similarly argues against adaptationism, and particularly against the view that the female orgasm is an adaptation. While it may seem obvious to many that female orgasm must be an evolutionary adaptation, she points out, this idea actually has no significant empirical support. There have been a number of theories advanced to explain just how the female orgasm contributes to the fitness of the woman or of the man with whom she mates. These include the theories that the orgasm reinforces the pair bond; that it prevents violent males from attacking the mate's offspring; or that sperm are preferentially sucked into the reproductive tract upon mating with genetically superior males. Lloyd examines each of these explanations and the evidence advanced for them, and discusses some of the biases that seem to be involved in evaluating these theories. In recent years, the most popular explanation has been the female choice theory, that females preferentially orgasm with 'high quality' males, which, unlike past theories, explains the segment of women who sometimes do and sometimes do not have an orgasm with intercourse. She criticizes the startlingly poor evidence for this theory, which contrasts sharply with the strong evidence for the theory that the female orgasm is a by product of selection on the male orgasm. It seems that some biologists approaching the question of the female orgasm have evaluated the evidence backwards: strong evidence for the by product hypothesis has been ignored, while exceptionally weak evidence in favour of the female choice theory has been elevated to the level of 'fact'. Lloyd concludes by responding to several objections by feminists.

In Chapter 8, which opens Part IV: Sexual Selection and Morality, Lucrecia Burges, Camilo J. Cela Conde and Marcos Nadal write that ethologists are interested in the evolution of altruism because it constitutes a paradox in the light of Darwinian theory. After arguing that explanations of altruism on the basis of natural selections do not quite deserve the acceptance they have received in some quarters, they explore the extent to which sexual selection may have played a role in the appearance of human moral traits. It has been suggested that because certain moral virtues, including altruism and kindness, are sexually attractive, their evolution could have been shaped by the process of sexual selection. Their review suggests that although it is possible that sexual selection played such a role, it is difficult to determine the extent of its relevance, the specific form of this influence, and its interplay with other evolutionary mechanisms.

Taking the bull by the horns, in Chapter 9 Julia Sandra Bernal uses biological and anthropological data to explain how sexual selection might have influenced the evolution of morality and even that of the law. The evolution of morality and law, she argues, can be explained by the socialization of hominids through recurrent cooperative reciprocal interactions based on mutual interdependency and equality. These interactions resulted in the greater biological efficiency of gregarious individuals when compared to non-cooperative or solitary individuals. One of the factors that could have increased the biological efficiency of our ancestors was initially the mutual sexual selection between males and females due to their shared interest in the survival of their offspring, which led, among other things, to the selection of a mating system with a greater tendency towards monogamy than that of our closest primate relatives. *Efficient* behaviours that allowed for the survival of our ancestors were enhanced through psychological reinforcement mechanisms such as social approval, abstracted as mandatory behavioural norms. Contrary behaviours, by contrast, were sanctioned, banned or inhibited through persuasion, discrimination or punishment. Norms are, then, the expression of a coordination phenomenon that once acted as a mechanism of social balance that limited the power held by any one individual while controlling for cheaters and spongers. Even at the level of highly stratified societies, as the anthropological data suggest, there are strong reasons to seek consistency between the system of law and such a mechanism.

In Part V: Sex, Reproduction and Evolutionary Psychology, William M. Brown bases his conclusions about the importance of symmetry, as a sign of developmental stability, in sexual selection on a variety of experiments, including his own. Fluctuating asymmetry (FA) – random departures from perfect symmetry – is an inverse measure of developmental stability: an organism's ability to reach an adaptive endpoint despite environmental perturbations. This chapter reviews the evidence that low FA individuals are more likely to survive the rigours of natural selection and are oftentimes selected as mates. Despite the success of the approach, controversies remain. Indeed, it is not clear why some organisms have better buffering capacities than others. This chapter introduces a new genetic co-adaptation theory for developmental stability called *genomic antagonism reduction*. Simply stated, when intragenomic and intergenomic conflicts are minimized, individuals will be better at buffering ontogenetic stressors. Some previous findings are reviewed and determined to be consistent with this proposition. For example presence of an ultra-selfish genetic element (i.e., the t-allele) increases fluctuating asymmetry in house mice. Brown presents new analyses from previously published datasets suggesting that socially monogamous and polyandrous avian species have lower FA than lekking polygynous avian species. This finding is consistent with the proposition that polyandry is an evolved strategy to avoid genomic incompatibles caused by selfish genetic elements. Previous debates regarding the generality of the negative relationship between heterozygosity and FA may also be resolved in part by a *genomic antagonism reduction* perspective. Specifically, in species with reduced likelihood of intragenomic conflict, developmental stability may be more achievable. The future of this approach for the study of developmental stability is discussed.

A different line of psychological and neurophysiological experimentation, however, has led to the conclusion that sexual attraction involves some important deviations from symmetry that reflect, for example, hormonal influences on fecundity. The most prominent practitioner of this line of research is perhaps Victor S. Johnston, whose chapter, using both theoretical and empirical research, will offer some novel insights into the evolution of human sexual aesthetics and the design of the human mind. Using facial beauty as an example of a sexually selected trait, Johnston's chapter examines both the biological factors that underlie attractive facial traits and their interaction with the observer's brain.

Part VI: Eugenics from Natural to Social Selection, the final section of the volume, explores the leap from natural to social selection and takes up eugenics, a doctrine that shaped views on reproduction and sex stretching to the present day. The theory of selective human breeding is inextricably bound to the fate of evolutionary ideas and, because it provided the rationale for the most devastating annihilation of human lives in the twentieth century, eugenics has been conspicuously used to discredit the Darwinian legacy. The relation between the two strands of thinking is a complex one for they are antithetical but share points of contact. Eugenics is prescriptive and purports to manipulate evolution through a joint process of social and biological engineering, while Darwin's theory is purely descriptive and premises evolution on natural selection. Yet both were steered by the desire to understand the process of evolution and the underlying mechanisms of heredity and are reliant upon a common intellectual ancestor, the Malthusian view on population.

In Chapter 12, Eve-Marie Engels describes the common ancestry and parting of the ways, debunking the myth that conflates Darwinism with eugenics. She exposes misunderstandings created by the additional slippage of Social Darwinism and outlines the conceptual and historical markers that differentiate each theory. Scrutiny of chronology, the intellectual tenor, values, personal circumstances and the social background that shaped these strands of thinking make a compelling strategy to dispel assumptions ingrained in academia and public discourse. With the benefit of intimate knowledge of primary sources, Engels points to passages in Darwin's works that have been used in support of particular agendas and calls for a different interpretation, an interpretation informed by contextual issues including Victorian debates on contraception, as well as by Darwin's strong moral sense and concern for all of humanity.

Finally, in Chapter 13, Filomena de Sousa looks at the origins of eugenics in Britain and ramifications in the United States and seeks to bring a historical perspective to bear on current discussions on reproductive genetics and biotechnology. She examines the formation of a eugenics web that brought together leading scientists, intellectuals and reformers that sought to direct the evolution of the human species towards a social order based on nativism entailing biological hierarchy. The chapter surveys recurrent themes in eugenics that emerged from debates on reproduction, sex, evolution and heredity brought to the fore once more with the second revolution in genetics. Genomic knowledge opened new possibilities for the way our species reproduce; but ranking reproductive worth, as de Sousa argues, is a source of anxieties that remains unsolved.

It is our hope as editors that our readers will take the contributions to this volume as a platform from which to launch further investigations of their own about these fascinating subjects. We ourselves are struck by the many points at which apparently different approaches and different questions about sex, reproduction and evolution make intellectual contact and throw light on each other. If this volume manages to convey that sense of intellectual stimulation, we will feel very amply rewarded.

1 FROM BIRTH TO DEATH: THE EVOLUTION OF LIFE HISTORIES IN GUPPIES (*POECILIA RETICULATA*)

David. N. Reznick

Introduction to Life History Evolution

The Life History Dilemma

I study the evolution of the entirety of the life history, from birth to death, in sexually reproducing organisms. I define the life history as the composite of all of the traits that contribute to how an organism propagates itself. The most important components of the life history in the animals I study are the age at maturity, how often the animal reproduces after attaining maturity, and how it divides its resources among growth, maintenance, storage (e.g., fat reserves) and reproduction.

My interest in life histories is motivated by the theoretical importance of life history traits and by the natural history of life histories. On the theory front, the components of the life history are very closely allied to how we define Darwinian fitness. Fitness equals the relative success of different phenotypes in contributing offspring to the next generation. Life histories are defined by the timing and quantity of resources that are devoted to producing offspring. The compelling feature of the natural history of life histories is that they are so incredibly variable among species and populations. In plants and animals, the age at first reproduction can vary from days to centuries and the frequency of reproduction can vary from once (as in a salmon or some agaves) to dozens of times or more. The number of offspring produced during the lifetime of a parent can vary from a few, as in humans or elephants, to hundreds of millions, as in the Ocean Sunfish (*Mola mola*), with a commensurate difference in the resources devoted to each baby. The general question addressed in studies of life history evolution is thus 'if the components of the life history are such important components of fitness, then what can explain the evolution of the remarkable diversity of life histories that we see in nature'?

A Brief History of Life History Theory

What did Darwin have to say on the topic? Even though Darwin was the founder of evolution and even of many prominent subdisciplines allied with evolution, he had little to say about life histories. In Chapter 3 of *On the Origin of Species* (Struggle for Existence), Darwin contemplated why some organisms produce few eggs and others produced many.

> A large number of egg production is of some importance to those species, which depend on a rapidly fluctuating amount of food, for it allows them rapidly to increase in number. However, the real importance of a large number of eggs or seeds is to make up for much destruction at some period of life; and this period in the great majority of cases is an early one. If an animal can in any way protect its own eggs or young, a small number may be produced, and yet the average stock be fully kept up; but if many eggs or young are destroyed, many must be produced, or the species will become extinct.[1]

Darwin had little more to say on the subject. In fact, he argued that the main consequence of evolution by natural selection was that it selected for a longer life span. If an animal lived longer then it had more opportunities to reproduce. Darwin distinguished between natural selection and sexual selection by arguing that, while natural selection acts on survival, sexual selection acts on reproductive success. He further argued that since survival is so much more important than reproductive success, natural selection will generally dominate sexual selection in shaping how organisms evolve.

R. A. Fisher (1930) and the Modern Synthesis

While *On the Origin of Species* did much to convince the world of the importance of evolution, it was far less successful in promoting natural selection as the mechanism that causes evolution. Our current interest in natural selection, and the adaptations it causes, dates to the modern synthesis that began in the 1930s. This is also when R. A. Fisher took the first step in generating interest about life history evolution. He acknowledged Darwin's observation that the vast overproduction of offspring is an adaptation to the risk that each of them faces. Fisher then went on to say:

> It would be instructive to know not only by what physiological mechanisms a just apportionment is made between the nutriment devoted to the gonads and that devoted to the rest of the parental organism, but also what circumstances in the life-history and environment would render profitable the diversion of a greater or lesser share of the available resources towards reproduction.[2]

Fisher thus widened the field of inquiry about life histories to include a consideration of how organisms allocate resources. He took the critical step of viewing the production of offspring as an investment of finite resources in reproduction, as opposed to the parent.

Fisher also formalized a way of looking at allocation to reproduction by deriving a function for reproductive value, which is the expected contribution of offspring to the next generation as a function of age. An individual's reproductive value increases from birth to sexual maturity. The reason for this increase is that not all newborn individuals will survive to reproductive age. As an individual approaches reproductive age, then the probability of surviving to reproduce, and hence its value in terms of what it will contribute to the next generation, increases then peaks at first reproduction. After maturity, value declines because some offspring have already been produced, which diminishes the expectation of future reproduction. In addition, after maturity the mortality rate will increase as a consequence of senescence.

George Williams[3] and the Partitioning of Reproductive Value

George Williams took the next step by dividing reproductive value into two components – reproductive effort versus residual reproductive value. Reproductive effort is the component of reproductive value that is at stake in the offspring that are being produced right now. Effort is a function of how many offspring are produced and how much is invested in each of them. Residual reproductive value is the expectation of future reproduction.

Williams argued that there would be a trade off between reproductive effort and residual reproductive value. Investing more now will in some way detract from what can be invested in the future because resources are finite. Your investment in reproduction now must take away from something like growth, maintenance or storage, all of which is expected to detract from the future potential to reproduce. Growth is important in some organisms because the number of offspring they can reproduce is directly proportional to size. Investing more in reproduction now and reducing growth will thus reduce how many offspring can be produced in the future. If investing more in offspring means reducing investment in storage or maintenance, then the consequence may be reducing the odds of surviving to reproduce again.

Williams built his argument by imagining a bird rearing a group of hatchlings still in the nest. Each day, the bird faces the repeated question 'should I venture out once more to catch an insect to feed my offspring, investing time, energy and risking being caught by a predator, or instead call it a day?' If the answer is 'yes', then it is accompanied by the benefit to the babies of getting more to eat and the cost to the parent of the effort and risk invested in getting the food. If the answer is 'no' then it incurs the cost of not feeding the babies again. Williams solved for the condition where the net benefit of answering 'yes' exactly equalled the net cost of answering 'no'. The important feature of this approach is that the answer is directly proportional to the immediate benefit of the added investment, but inversely proportional to the residual reproductive value, or the expectation of future reproductive success. Doing more now means taking away from the future. This simple elaboration on Fisher's concept of reproductive value and formalization of the concept of tradeoffs between now and later enabled us to envision how to address questions about how the allocation of resources to reproduction should be shaped by natural selection.

Demographic Theory

Subsequent life history theory proposed what the optimal allocation of resources across time and development should be, based upon assumed tradeoffs among the different ways in which organisms can allocate finite resources. For reasons that will become clear below, I gravitated to theory that modelled the evolution of optimal resource allocation in response to externally imposed risks of mortality.

A common feature of the lives of all organisms is that they face a constant risk of mortality caused by factors like predators or disease. One body of theory, initiated by M. Gadgil and P. W. Bossert,[4] developed models that predicted how life histories should evolve in response to an increase in the risk of mortality that was imposed selectively on juveniles versus adults, since the risk of death and the ability to reproduce vary with the age of the organism. One prediction that accords well with common sense is that an increase in the risk of mortality in adults, perhaps imposed by a predator that preys selectively on large prey, will select for individuals that begin to reproduce at an earlier age and have higher reproductive effort, or a higher rate of allocation of resources to reproduction. Both of these changes (beginning to reproduce earlier and investing more in reproduction) should carry a cost in the form of reduced survivorship or a reduced capacity to reproduce in the future; however, the magnitude of this cost is reduced by the fact that the expectation of survival is low regardless of how resources are allocated because of the high risk of mortality imposed by the environment (i.e., by predators or disease). These same ideas were elaborated upon by others who followed.[5]

Life History Evolution in Guppies

Introduction to Guppies

Guppies (*Poecilia reticualta*) are small, live-bearing fish from north-eastern South America and neighbouring islands. Adult males are 13–17-mm long and adult females are 14–>30-mm long. They have a short generation time, with the time between birth and when a female gives birth to her first litter of babies being as short as ten weeks. Females will give birth to a new litter of live-born young every three to four weeks thereafter. The guppies that I study come from the Northern Range Mountains of Trinidad.

The Northern Range Mountains of Trinidad offer a natural laboratory for studying life history evolution. These mountains are covered by seasonal tropical rain forest, grading into cloud forest at higher elevations. They receive three to four metres of rain per year. The rivers draining these mountains flow over steep gradients punctuated by waterfalls. The waterfalls often define the boundary between different fish communities. The most diverse communities are found downstream, where groups of streams converge to form larger rivers. Species diversity decreases upstream because waterfalls block the upstream dispersal of some species. On the south slope, dramatic waterfalls can exclude all species save one, the killifish *Rivulus hartii*, which is capable of overland dispersal. On the North Slope, these simplest of communities also include a small goby that is capable of moving upstream through the steepest of barriers.[6]

The succession of communities, from diverse communities downstream to progressively simpler communities upstream, is repeated in many, parallel drainages. These parallel drainages provide us with natural replicates. Guppies are found in most of these communities. In the complex, downstream communities they cohabit with a diversity of predators. Some of these predators prey on adult-size classes of guppies. I will refer to these as 'high predation' communities. Waterfalls often exclude predators but not guppies, so guppies found above waterfalls have greatly reduced risks of predation and increased life expectancy. At the extreme, guppies are found only with the killifish *Rivulus hartii* that is capable of overland dispersal. *Rivulus* feeds predominantly on invertebrates, but sometimes also eats guppies. When it does, it feeds predominantly on smaller, immature-size classes of guppies. I will refer to these as 'low predation' communities.[7]

Mark-recapture studies on natural populations revealed that guppies from high predation localities experience substantially higher mortality rates than guppies from low predation localities.[8] Life history theory predicts that these differences among high predation and low predation communities in the risk of mortality will select for differences in guppy life histories. In high predation communities, the expectation is that higher adult mortality rates will select for those individuals that attain maturity at an earlier age and have higher levels of reproductive effort than seen in guppies from low predation environments.

I tested this prediction by first comparing guppies from high and low predation communities in a number of different rivers in which both community types were represented. Other researchers have analysed the patterns of genetic variation among these populations and have shown that each river indeed represents an independent event in which guppies have adapted to the alternatives of high versus low predation environments.[9] A likely pattern of colonization is for guppies from high predation localities downstream to have invaded previously guppy free headwater streams, then adapted to life without predators.

I used two methodologies for comparing the life histories of guppies from high and low predation environments. The first involved collecting and preserving fish from a series of paired high and low predation localities, preserving them, and then dissecting them to characterize their life histories. The second was to collect wild-caught females from paired high and low predation localities and bring them back to the laboratory. I isolated each wild-caught female in her own aquarium, and then exploited a convenient feature of guppy repro-

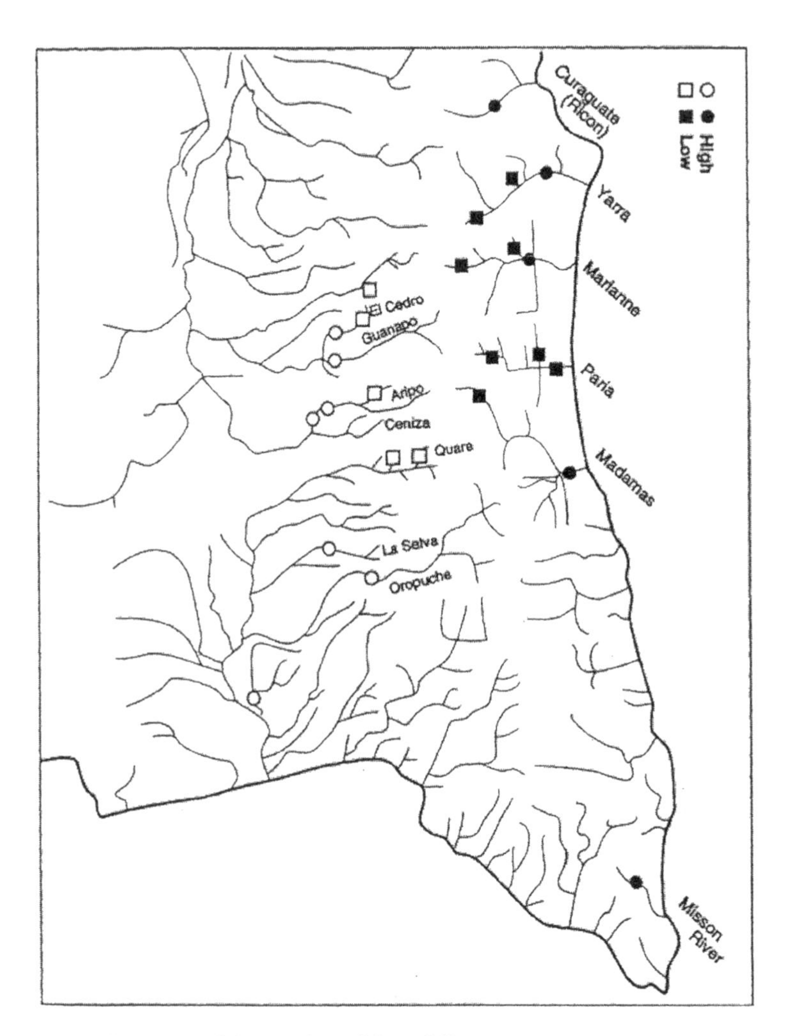

Figure 1.1: Illustration of the sampling of three different community types in ten rivers of the Northern Range Mountains of Trinidad. Source: D. N. Reznick, H. F. Rodd and M. Cardenas, 'Life History Evolution in Guppies (*Poecilia reticulata*: Poeciliidae). 4. Convergence in Life History Phenotypes', *American Naturalist*, 147 (1996), pp. 319–38.

The three community types are those where guppies co-occur with a diversity of predators (high predation), those where they co-occur with just the killifish *Rivulus hartii* (low predation) and those where *Rivulus hartii* is the only fish species present. In this particular study, we sampled from five drainages each on the north and south slopes of the Northern Range Mountains. On the south slope, the predators are a subset of those that would be found on the mainland of South America because all of these rivers were once linked to rivers on the mainland. The most common predators are cichlids and characins. The rivers on the north slope were never linked to the mainland, so all of these species are absent. The predators are instead ones that colonized from the ocean, including gobies and mullets. Even though the predator faunas are different, the impact of the predators on the evolution of guppy life histories is the same, which argues for the generality of predator-induced mortality as an important agent of selection for guppy life history evolution.

ductive biology. Female guppies store sperm and reproduce continuously. Each isolated female produced a succession of litters of live-born young. I reared the offspring from each female in single sex groups until they attained maturity, then set up crosses between individual males and females from different pedigrees. I included one male and one female offspring from each wild-caught female, but each cross represented a different combination of offspring from wild-caught females. The design enables me to equally represent all wild-caught females and retain as much genetic diversity as possible from the original wild-caught mothers. I then reared the offspring from these crosses, which were now the grandchildren of wild caught females, one per aquarium on quantified rations. In these conditions I can quantify the entirety of the life history, including the age and size at first reproduction, how often the fish reproduced, the number of offspring produced and the size of individual offspring. Because I quantified food availability, I was also able to estimate reproductive effort as the percentage of consumed food that was devoted to offspring. Because the fish that had been reared for two generations in a common environment, it was safe to assume that any differences among populations in life histories observed had a genetic basis.

The first type of comparison, among populations of fish that had been collected in the wild, preserved and dissected, had the virtue that it enabled me to compare guppies from high and low predation communities from a large number of rivers. It had the cost that I was only able to characterize part of the life history. I could quantify the proportion of body weight that consists of developing offspring, the number of offspring per litter and the size of individual offspring, but I could not quantify the age at maturity or frequency of reproduction. The second type of comparison, among the laboratory reared grandchildren of wild-caught females, enabled me to quantify the entire life history, but I was only able to work with four to six populations at a time and it took a year to complete the study. Both types of study yielded the same result, hence here I only summarize the results of one of my laboratory experiments (see Table 1.1).

Table 1.1: Early life histories for the second laboratory reared generation of laboratory reared fish from two rivers in Trinidad.

Drainage	Predation	Age at first Parturition (days)	Age at third parturition (days)	Interbrood Interval (days)	Offspring Dry Mass (Litter 1 – mg)	Number of offspring in the first three litters
Oropuche	High	90.6 (2.7)	142.2 (2.9)	25.7 (0.38)	0.93 (0.03)	17.9 (1.03)
	Low	97.0 (2.1)	153.1 (2.7)	28.3 (0.7)	1.29 (0.03)	15.6 (0.9)
Yarra	High	82.2 (1.9)	133.4 (2.4)	25.2 (0.6)	0.73 (0.02)	21.9 (1.5)
	Low	92.0 (1.6)	150.4 (3.0)	29.1 (1.07)	1.04 (0.02)	14.8 (1.0)

Source: D. N. Reznick, M. J. Bryant, D. Roff, C. K. Ghalambor and D. E. Ghalambor, 'Effect of Extrinsic Mortality on the Evolution of Senescence in Guppies', *Nature*, 431 (2004), pp. 1095–9; and D. N. Reznick, M. J. Bryant and D. Holmes, 'The Evolution of Senescence and Post-Reproductive Lifespan in Guppies (*Poecilia reticulata*)', *PLOS Biology*, 4 (2006), pp. 136–43.

The Yarra River is from the northern slope of the Northern Range Mountains. The Oropuche River is from the south-eastern slope of the Northern Range Mountains. The guppies from these two rivers are genetically quite distinct from one another and represent independent occurrences of guppies adapting to high versus low predation environments. You will see a strong effect of drainage of origin, which is statistically significant for every variable, but you will also see consistent life history differences between the guppies derived from high and low predation environments in each drainage. While these differences may seem subtle, they sum to vast differences in estimated population growth rate and do so because of their cumulative nature. Guppies from high predation localities are 7–11 per cent younger at first reproduction, reproduce more often and give birth to more and smaller offspring.

The results of these comparisons, which have now been repeated in six different drainages on the north and south slopes of the Northern Range Mountains, consistently support the predictions of life history theory. Guppies from high predation communities are younger and smaller at maturity than their counterparts from low predation communities. They also reproduce more often and devote more resources to each litter of young. The combined consequence of more frequent reproduction and investing more in each litter also means that they have higher reproductive efforts than do guppies from low predation environments.

These results establish a consistent correlation between predation, risk of mortality, life histories and predictions from theory, but fall short of establishing a cause and effect relationship between predation and life history evolution. The next step was to perform selection experiments in nature. Rivers can be treated like giant test tubes, since fish can be introduced into portions of stream bracketed by waterfalls, creating *in situ* experiments.[10] When guppies were transplanted from high predation environments below a barrier waterfall to previously guppy-free environments above a waterfall, they evolved life histories that matched those normally seen in low predation environments. Some features of the life history showed significant life history evolution in as little as four years (Reznick and Bryga (1987); Reznick, Bryga and Endler (1990); Reznick et al. (1997)).[11] In a different experiment, we transplanted predators over a barrier waterfall that had previously excluded predators but not guppies. Here I observed the evolution of earlier maturation in males and females after only five years. These results argue that the presence or absence of predators imposes intense selection on the evolution of guppy life histories. These experiments also represent a rare instance in which it has been possible to test some aspect of evolutionary theory with experiments done on natural populations and to measure the rate of evolution in a constructed episode of directional selection. The rates that we observed were on the order of ten thousand to ten million times faster than had been seen and thought to be rapid evolution in the fossil record.

Table 1.2: Results from experimental studies of evolution in which guppies from a high predation locality were transplanted into previously guppy-free localities above barrier waterfalls that excluded all species of fish except *Rivulus hartii*.

Measurement	1 σ means (standard error) Control	Experiment	Response R	Rate (10^3 darwins)
	Aripo River (11 years or 18.1 generations)			
Male age (days)	48.6 (1.1)	58.2 (1.3)	9.6*	16.4
Male size (mg)	67.5 (1.6)	76.1 (1.9)	8.6*	10.9
Female age (days)	85.6 (2.2)	93.5 (2.6)	7.8*	8.0
Female size (mg)	162.3 (6.4)	189.2 (7.4)	26.8*	13.9
	El Cedro River (4 years or 6.9 generations)			
Male age (days)	60.6 (1.8)	72.7 (1.8)	12.1*	45.0
Male size (mg)	56.0 (1.4)	62.4 (1.5)	6.4*	27.1
Female age (days)	94.1 (1.8)	95.5 (1.8)	1.4 (ns)	3.7
Female size (mg)	116.5 (3.7)	118.9 (3.7)	2.4 (ns)	5.1
	El Cedro River (7.5 years or 12.7 generations)			
Male age (days)	47.3 (1.1)	52.5 (0.6)	4.9	13.9
Male size (mg)	71.5 (1.1)	74.4 (0.7)	2.9*	5.3
Female age (days)	75.8 (1.8)	80.4 (1.0)	4.6†	7.9
Female size (mg)	141.8 (5.1)	152.1 (3.2)	10.3*	9.3

* $P < 0.05$. †P $0.05 < P < 0.10$.

Source: D. N. Reznick, F. H. Shaw, R. H. Rodd and R. G. Shaw, 'Evaluation of the Rate of Evolution in Natural Populations of Guppies (*Poecilia reticulata*)', *Science*, 275 (1997), pp. 1934–7.

The data included here are for male age and size at maturity and female age and size at first parturition. The data were collected on the second generation of laboratory born fish derived from wild-caught females, following the same protocol as for the fish represented in Table 1.1 and Figure 1.3. 'Control' fish are the second generation descendents of adult females collected from the high predation locality that the introduced fish were collected from. 'Experimental' fish are the second generation descendants of adult females collected from the introduction site. The differences between populations observed under such conditions can be interpreted as having a genetic basis and thus provide evidence that evolution has occurred. The means are least square means derived from the statistical comparisons of these populations. Sample sizes for the control in the Aripo River were 30 males and 29 females; for the experimentals they were 24 males and 22 females. In the El Cedro river, 4-year assay (meaning that it was based on wild caught females collected four years after the introduction), the control included 44 males and 43 females; the experimental included 42 males and 43 females. In the El Cedro 7.5 year assay, the control included 36 males and 40 females; the experimental included 129 males and 110 females. 'Response' is the estimated response to selection, or the difference between the control and experimental values. The statistical significance of the response is based on an analysis of variance. Significance is based on one-tailed probabilities because the comparative studies provided a basis for predicting how the populations would evolve in response to the introduction. 'Rate' is the estimated rate of change in darwins, calculated as (lnX2 – lnX1)/ delta t, where X1 and X2 are the values of the trait at the beginning and end of the time interval, respectively, and delta t is the length of the interval in years. We estimated X1 and X2 using the values of the traits for the control and introduction populations, respectively. Gingerich(1983) reports that artificial selection experiments attained values of 12,000 to 200,000 darwins, with a geometric mean of 58,000 darwins. In contrast, the geometric mean rates for the fossil record range from 0.7 to 3.7 darwins, with the estimated rate being inversely proportional to the time interval over which it was estimated. Our observed rates in guppies are often of the same order of magnitude as the geometric mean rate for artificial selection, and four to seven orders of magnitude higher than seen in the fossil record.

III. From Birth to Death

Some of the most important results are the ones that do not fit expectations. Here I offer a synopsis of two results that have told us that age-specific predation alone cannot explain how guppy life histories have evolved. Along the way, they offer more general messages about the likely complexity of any adaptation we see in natural populations.

From Birth ...

Life history theory often predicts the proportion of resources that are devoted to reproduction, versus other things, without specifying how these resources are to be allocated to offspring. Whether it is best to produce many small offspring or few large ones is often addressed with a different body of theory that deals explicitly with the trade-off between the number of offspring produced and the quality of individual offspring. An interesting feature of guppy life history evolution is that the evolution of reduced reproductive effort in guppies from low predation environments is consistently associated with the evolution of the production of fewer, larger offspring. Conversely, the evolution of high reproductive effort in guppies from high predation environments is associated with the evolution of the production of many, small offspring. Why is there such a consistent trend? The associated demographic theory that was so successful in predicting the evolution of the age at maturity and reproductive effort is silent on the issue of how reproductive effort is apportioned, so we had to address the issue of the evolution of offspring size in some other way.

First, I must digress and describe some work on the comparative ecology of guppies from high and low predation environments to develop the alternative hypotheses for the evolution of offspring size. We executed a study of the comparative ecology of guppies from high and low predation communities and found that there were consistent differences between them.[12] We chose communities that tended to be in close proximity to one another, separated by barrier waterfalls that excluded predators, so that the differences in physical environment were minimized. We found that guppies live at higher population densities (four to five fold higher, when estimated as the biomass of guppies per unit area or volume of the stream). The populations also differ dramatically in age/size structure. In high predation localities, guppy populations are dominated by small, young fish, while in low predation populations the age and size distribution is more evenly distributed, with a proportionately much higher representation of older, larger fish. These differences are a consequence of mortality and birth rates and parallel what we see in human populations. Guppies in high predation localities experience high birth and death rates while those from low predation localities experience low birth and death rates.

We also found that individual guppies from high predation environments have higher growth rates. The best explanation for this difference is that food availability is higher in high predation environments.

The explanation most consistent with these observations is that they reflect the indirect effects of predators. Predators eat guppies, but at the same time they reduce the population density of guppies. As a consequence, guppies that survive have higher resource availability. This kind of indirect effect of predators on the abundance of prey is well known from a diversity of other communities; for example, the extirpation of predators from the northeastern United States has resulted in an explosion in the deer population.

This correlation between the risks of predation and indirect effects of predators leads to a hypothesis for why guppies from low predation localities produce larger offspring. Perhaps larger offspring have higher fitness, perhaps because they are superior competitors, when food is scarce.

The first general theory for the evolution of offspring size was presented by C. C. Smith and D. Fretwell in 1974.[13] They argued that there is an optimal egg size for a mother to produce that maximizes the mother's fitness. Larger offspring are presumed to have some fitness advantage over smaller offspring, but to be associated with the cost to the mother of reducing the number of babies she can produce. Their theory balances the gain of producing bigger eggs against the loss of producing fewer eggs.

C, Jorgenssen et al. revisited Smith and Fretwell's general model, but made it more concrete by modelling the evolution of offspring size as a joint response to offspring risk of mortality and resource availability.[14] When mortality rate alone is included in the model, the prediction is that the organism will produce larger offspring. The reason is that such offspring are presumably better able to escape predators and will take less time to grow beyond the small size class that is susceptible to predation. In fact, it has been empirically observed in many organisms that larger offspring have lower mortality rates. If resource availability alone varies, then high resource availability will favour the evolution of smaller offspring size, since abundant food will sustain high growth rates and compensate for small birth size. They incorporated a diversity of other factors in their model, such as a consideration of the consequences of extended parental care, which reduces the number of offspring that can be produced while at the same time improving their probability of survival.

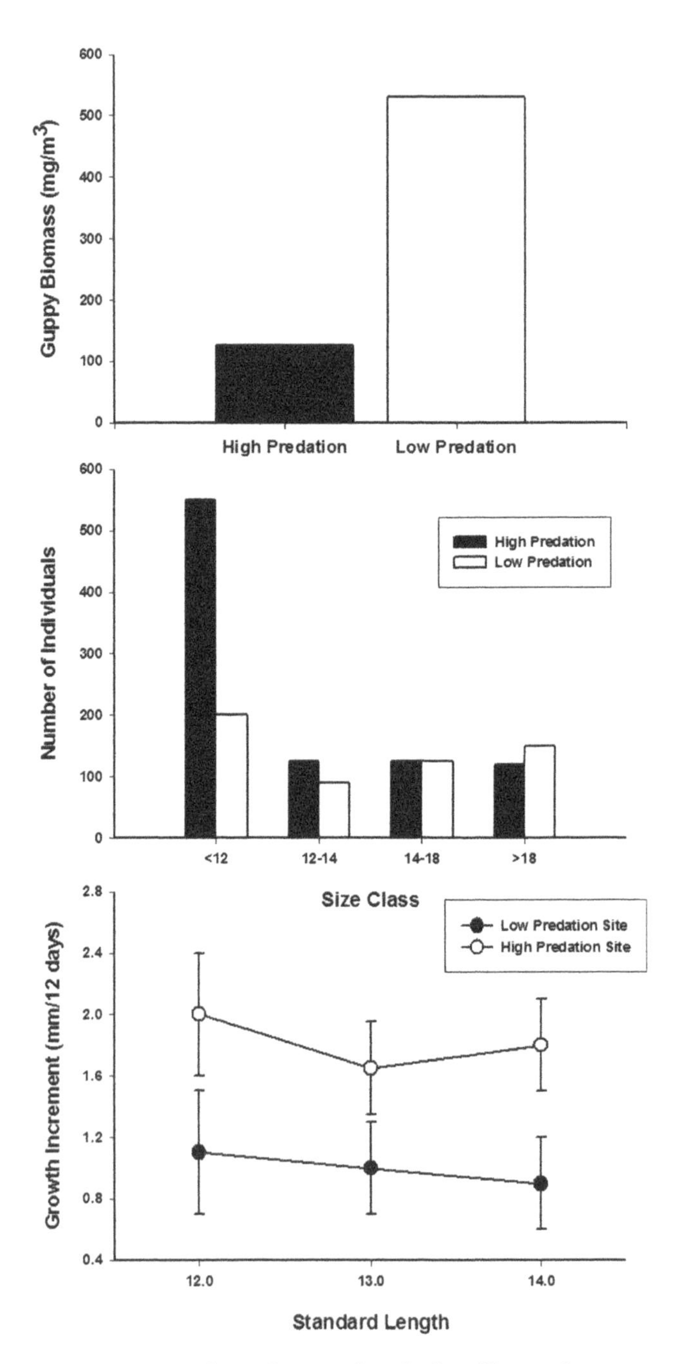

Figure 1.2: Comparative ecology of guppies from high and low predation environments. Source: D. N. Reznick, M. J. Butler IV and F. H. Rodd, 'Life History Evolution in Guppies 7: The Comparative Ecology of High and Low Predation Environments', *American Naturalist*, 157 (2001), pp. 126–40.

Six low and six high predation localities were represented in these samples. (a) The biomass of guppies per unit volume of water, illustrating the much higher population densities of guppies in low predation environments; (b) The size distribution of guppies from these same sizes, illustrating the predominance of small, young guppies in high predation environments and the more even size distribution of guppies from low predation environments; (c) Growth rates of small, immature female guppies from these same high and low predation environments, illustrating the higher growth rates of guppies from high predation environments. Note that there were no temperature differences among these sites, which is important because temperature can influence growth rate.

Jorgenssen et al. chose guppies as a special case for testing their model.[15] First they modelled the evolution of offspring size in response to risk of mortality alone. Mortality rate declines with egg size, but making bigger eggs means producing fewer of them. If the mortality rate of juveniles alone shapes the evolution of offspring size, then the prediction is that a higher risk of juvenile mortality will select for increased size at birth. Being smaller at birth means being less able to escape a predator and taking longer to exceed some minimum size that might be susceptible to predation. If offspring size evolved in response to mortality risk alone, then the prediction is that high predation guppies should produce larger offspring. However, we also know that guppies from high predation environments have higher growth rates than those from low predation environments. When the expected slower growth rate of offspring in low predation environments is added to the model, then it predicts that the low predation guppies should instead produce larger offspring. Theory thus suggests that differences in growth rate, and presumably food availability, are necessary to explain the evolution of larger offspring size in low predation localities.

Farrah Bashey performed two kinds of experiments to evaluate the consequences of size at birth in guppies.[16] One experiment was performed on populations of guppies reared in large, pond-like enclosures in a greenhouse. The other was performed on pairs of guppies reared together in a two-gallon aquarium in the laboratory. In the pond experiment, she found that larger offspring had higher growth rates than smaller offspring, but only when reared at high population densities. When they were reared at low population densities the growth rates were the same. In paired competition studies in the laboratory, she showed that larger offspring were younger and larger at maturity than smaller offspring when they were reared on low food availability. These differences disappeared when they were reared on high food availability. The two experiments suggest that large offspring have a competitive advantage and higher fitness when food is scarce, but the advantage disappears when food is abundant.

The combination of Jorgenssen's theory and Bashey's experiments argue that low predation guppies produce larger babies because they are born in to an environment where food is less abundant and competition is more severe. The comparative ecology of high and low predation environments reveals that guppies in low predation environments live at higher population densities and have lower growth rates. Theory plus these two types of empirical evidence thus suggest that part of what governs the evolution of different life histories in guppies from high and low predation environments is differences in population density and resource abundance. These differences are in turn potentially attributable to the indirect effects of predators on guppies, mediated through their impact on guppy population density and food availability.

... to Death

Early in his career, George Williams[17] applied the same logic of there being trade-offs among different components of the life history to explain the evolution of senescence. He proposed that senescence evolves as a by product of the evolution of the early life history via antagonistic pleiotropy. When natural selection causes the evolution of changes in reproductive investment early in life, the same genes that cause these changes have pleiotropic effects on late life performance. If the consequence of these genes is to cause increased allocation to reproduction early in life, via earlier maturity and/or increased allocation of resources to reproduction, then there should be a correlated decline in investment late in life and hence an acceleration in death rate at an earlier age and the correlated evolution of a shorter lifespan. If, conversely, there is selection in favour of genes that cause reduced reproductive investment early in life, then there should be a corresponding increase in the age of onset of senescence and a correlated increase in lifespan. If we apply this hypothesis to guppies, then it follows that guppies from high predation environments should initiate senescence at an earlier age and have shorter lifespans than their counterparts from low predation environments. These differences in the life history late in life are predicted to evolve as a consequence of how high and low mortality rates, as imposed by predators, cause the evolution of the early life history.

Before Williams, P. B. Medawar proposed an alternative theory for the evolution of senescence.[18] Medawar also argued that senescence evolves as a consequence of life expectancy, as determined by external features of the environment like predators, but via a different mechanism than that proposed by Williams. He noted that many mutations are deleterious and hence will be removed via natural selection. If the deleterious effects of a mutation are expressed early in life, before or shortly after sexual maturity, then selection against them will be intense. If they are expressed late in life, after reproduction has ceased, then there may be no loss of fitness associated with the mutation. Such late-acting mutations will tend to accumulate in the population. If an organism lives in an environment where live expectancy is short, then these late-acting mutations will accumulate more readily than they would in a population where life expectancy is longer. Senescence evolves as a consequence of this passive accumulation of deleterious mutations. If applied to guppies, then Medawar's theory makes the same prediction as Williams's theory, which is that guppies from low predation environments should have a delayed onset of senescence relative to their counterparts from high predation environments.

My colleagues and I sought to test this prediction with a laboratory experiment.[19] We initiated the experiment with the same protocol used for evaluating genetic differences among populations in life histories, which is to begin with wild-caught females, then to rear their offspring and grandchildren in a common, laboratory environment. We compared senescence among paired high and low predation localities from two drainages, the Yarra and the Oropuche. Prior genetic work had demonstrated that guppies from these two rivers are genetically distinct from one another and that each of them represents an independent occurrence of adaptation to life with and without predators. Including both rivers in the experiment thus tests the repeatability of life history evolution in response to predation. We reared the grandchildren, one per aquarium, with isolation beginning at an age of twenty-five to thirty days and ending at death. We characterized the life history of each individual with the age and size at first reproduction, age and size at all subsequent reproductive events, number of offspring at each reproductive event and age at death. Since we wanted to compare the life histories of these fish in environments that were free of extrinsic sources of mortality, but otherwise came as close as possible to replicating nature, we matched the water quality temperature and photoperiod to the prevailing conditions in nature.

L. Partridge and N. H. Barton have argued that when one compares the life histories of different populations or species in a laboratory common garden, then it is important that the conditions of the laboratory be an appropriate match to nature.[20] If the populations are from different environments and the lab environment matches one of them, but not the others, then that population may live longer as a consequence of being well adapted to that environment, rather than because it inherently lives longer. We knew *a priori* that guppies from high predation environments grow faster than those from low predation environments, probably because they have higher food availability. There are no consistent differences in the physical environments of these localities (for example, water temperature or chemistry). We reared the fish at high and low levels of food availability, chosen to result in growth rates that matched the average seen in guppies from high and low predation environments, respectively. Our high and low food treatments yielded asymptotic body sizes that approximated what we observed in high and low predation environments. Thus, the complete factorial experiment was a comparison of guppies from high and low mortality-rate environments in two drainages, for a total of four localities, crossed with two levels of food availability.

While theory is clear in making predictions about how senescence should evolve, it is less clear about how one should quantify senescence. We evaluated senescence as any age-specific decline in variables associated with individual fitness, specifically mortality, reproduction (estimated as the number of offspring produced per unit time) and physiological performance (estimated as the rate of acceleration and maximum swimming speed during an alarm response).

Our results were surprising.[21] Guppies from the high predation localities began to reproduce at an earlier age and produced more offspring early in life, as is usual for fish from high predation environments. However, they also have lower death rates and continued to have lower death rates until the very end of the experiment, when so few individuals remained alive that it was no longer possible to compare death rates with reasonable statistical confidence.

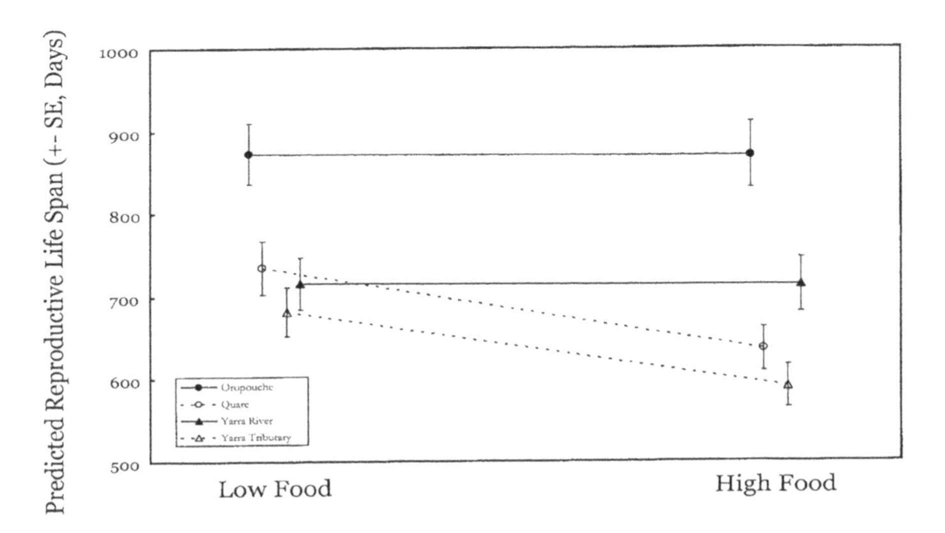

Figure 1.3a: Average reproductive lifespan of guppies from high and low predation environments in the Yarra and Oropuche Rivers. Source: D. N. Reznick, M. J. Bryant, D. Roff, C. K. Ghalambor and D. E. Ghalambor, 'Effect of Extrinsic Mortality on the Evolution of Senescence in Guppies', *Nature*, 431 (2004), pp. 1095–9; D. N. Reznick, M. J. Bryant and D. Holmes, 'The Evolution of Senescence and Post-Reproductive Lifespan in Guppies (*Poecilia reticulata*)', *PLOS Biology*, 4 (2006), pp. 136–43).

We also show the values for individuals reared on high versus low levels of food availability. Note that the guppies derived from the high predation environment have longer reproductive life spans than those from low predation environments, regardless of food availability.

All of the fish showed reproductive senescence, some in two different ways. The number of offspring produced per litter tapered off, then declined with age in virtually all individuals. Reproduction became irregular in many individuals. Early in life, they typically produced babies at regular intervals of approximately every twenty-five days. Later in life, some individuals seemed to skip litters, so that the interval of time between litters could be anywhere from 40 to >200 days. Many fish ceased reproduction entirely. However, the high predation fish continued to produce more offspring per unit time than low predation fish. The rate of decline in the rate of reproduction also tended to be slower in high predation guppies.

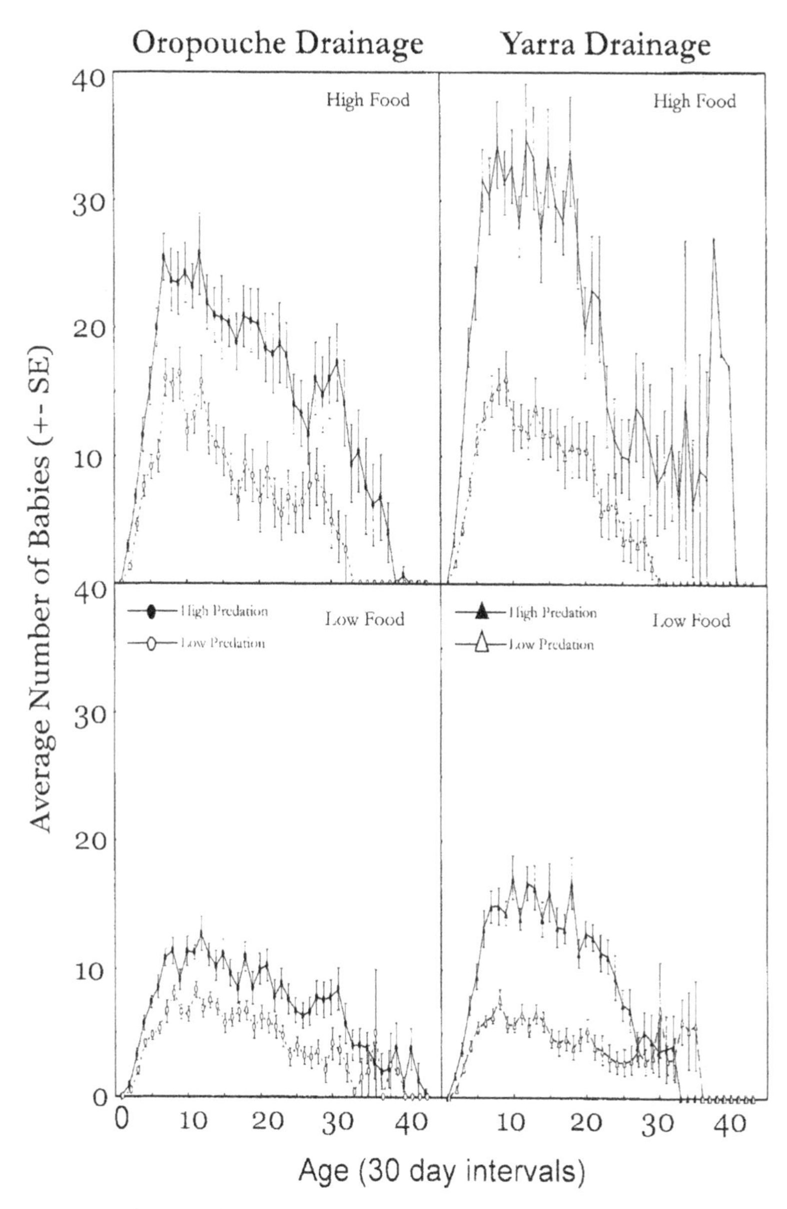

Figure 1.3b: Fecundity per month of the same fish as illustrated in Figure 3a, with values shown separately for fish reared on high versus low food rations.

All fish revealed reproductive senescence in the form of a peak then a decline in the rate of offspring production. However, the high predation guppies retained a reproductive advantage over low predation guppies throughout their lives and showed no evidence of a more rapid rate of decline in the rate of offspring production. In fact, the tendency was for guppies from high predation localities to be older when the decline began and for the rate of decline to be slower.

The only index of senescence that was consistent with Williams's[22] and Medawar's[23] predictions was the maximum rate of acceleration, which represents an index of neuromuscular performance. At twelve months, guppies from high predation localities were significantly faster than those from low predation localities. In a different study, we showed that the added acceleration gives them a significantly higher probability of surviving an attack from a predator. At twenty-six months, all guppies were slower, which is consistent with physiological senescence. Furthermore, the rate of acceleration declined more with age in guppies from high predation environments, so they were no longer significantly faster than those from low predation environments.

All of the comparisons between guppies from high and low predation environments were similar when they were reared on high or low levels of food availability, meaning that there was no statistical interaction between food availability and any aspect of the aging process. If the differences between guppies from high and low predation environments were in some way an adaptation to high and low food availability, respectively, then we expected an interaction, such that the differences between high and low predation guppies might be diminished, or even reversed at low food.

These results create what seems to be an impossible scenario. Guppies adapted to high predation environments invest more in reproduction early in life, yet seem to pay no price for it. In fact, they are superior to guppies adapted to low predation environments throughout their lives in terms of having higher rates of offspring production and lower mortality rates. If this is true in nature, then it argues that the high predation guppies are unconditionally superior and that the low predation life history should never evolve, yet it has evolved repeatedly when guppies were transplanted from high predation environments to low predation environments. How can we explain this result?

There are three results from theory that might help account for these results. The first came from Williams, who proposed that 'senescence should be more rapid in those organisms that do not increase markedly in fecundity after maturity than those that do show such an increase'.[24] If fecundity increases with age, then this increase can offset the age-specific decline in fitness caused by mortality. The rate of reproduction increases with age early in the life of all guppies because females continue to grow and fecundity is directly proportional to body size. However, the rate of increase in fecundity is higher in guppies from high predation environments. These guppies grow faster and attain larger asymptotic body sizes than those from low predation localities.[25] Their faster growth translates into a higher rate of increase in fecundity with age, which should offset some of the differences in mortality rate between high and low predation localities and hence result in smaller differences in senescence than expected from mortality rate alone.[26]

B. Charlesworth and P. Abrams show that the way senescence evolves can depend upon whether or not populations are subject to density regulation.[27] To paraphrase one prediction, predators increase mortality rate by eating prey, but may decrease mortality rates indirectly by reducing density and increasing per capita resource availability. If older age classes benefit more than younger age classes from higher resource availability, then higher mortality can cause the evolution of delayed senescence even though increased mortality without an indirect effect of density predicts the evolution of earlier senescence. We have already described how such indirect effects of predation occur in natural populations of guppies.[28]

Abrams and Williams and Day argue for a third possibility.[29] If the risk of being caught by a predator is dependent on the condition of the individual, then there will be positive selection for an improved ability to evade predators. A consequence of selection for improved escape performance will be deferred senescence in the age classes that have high reproductive potential. The strength of such selection should decline as an individual's reproductive potential declines with age. At the same time, the costs of deferred senescence will accumulate. Nevertheless, such condition-dependent selection can defer senescence to an older age than expected, based on Medawar's and Williams's predictions.

All three alternatives have properties that are applicable to guppies and all may contribute to our unexpected results. More generally, they challenge the simplicity of Medawar's[30] and Williams'[31] predictions and provide an incentive to consider the importance of some alternative models that have been proposed for the evolution of senescence.

In our most recent, as yet unpublished work, we have made some headway in addressing these alternatives. First, we have shown that natural populations of guppies do experience density regulation. Density regulation means that when the population has some carrying capacity, if the population is reduced below that capacity, then there will be some combination of increased birth rates or decreased mortality that causes the population to grow and return to its carrying capacity. If it exceeds its carrying capacity, then some combination of decreased birth rate and increased death rate will cause the population to decline and return to its carrying capacity. The key to evaluating Charlesworth's and Abrams's predictions is characterizing the demographic mechanisms of density regulation.[32] This work is in progress.

A more practical explanation for our unusual lab results is that the conditions under which we conducted the experiment in some way fail to represent circumstances encountered in nature. Specifically, we used food availability as a surrogate for the ecological differences between high and low predation environments. Food availability may be an imperfect surrogate for the consequences of high population densities in low predation environments versus low population densities in high predation environments. Our new experiments, conducted in replicate, artificial streams built alongside a natural stream in Trinidad, reveal that guppies from high predation environments do indeed have higher fitness than those from low predation environments, but only when they are compared at low population densities. These fitness differences disappear at high population densities. This result suggests that there is more to adaptation to population density than adapting to the correlated differences in food availability.

The exceptional qualities of the birth and death results tell us that predator-induced mortality alone cannot account for how guppy life histories have evolved. Indirect effects of predators, as mediated through their impact on guppy population density and resource availability, certainly matter.

How Generalizable are these Results?

There are different ways of asking whether or not all that is reported here is just an elaborate set of circumstances that applies to guppies or might instead represent results that are repeated throughout nature. One approach is to ask 'have other organisms been studied in the same way and are the results similar to those in guppies?' A second approach is to ask whether similar patterns emerge when we make statistical comparisons across an array of species, rather than among populations within a species. A third is to ask whether there is evidence from nature that suggests that the conditions faced by guppies in high versus low predation environments represent a kind of contrast that is repeated in other ecosystems. At all levels, I will argue that what we have seen in guppies is true for other species and in other ecosystems.

Within-Species Comparisons

First I will consider some detailed studies of intra-specific variation that are similar to my work on guppies. Jeff Dudycha and Alan Tessier performed a comparative study of the life histories of *Daphnia* (microcrustacea) from the *pulex* species complex.[33] They compared multiple populations of *D. pulex*, which live in temporary woodland pools, with *D. pulicaria*, which live in permanent lakes. These two species readily hybridize and produce viable, fertile offspring, hence their classification as being in the same species complex. The woodland ponds and lakes obviously differ in habitat duration, but also differ in productivity. The temporary ponds tend to be more productive than the lakes. Productivity in the lakes also varies considerably with season, being higher in the spring and autumn than in the summer and winter. Finally, there are differences in the types of predators found in each habitat, but the impact of the predators on the size-specific risks of mortality in the *Daphnia* are not known. There are thus multiple, confounded differences among the two habitat types. In a laboratory comparison of life histories, Dudycha and Tessier found that *D. pulex*, from woodland pools, are younger at maturity, produce more offspring early in life and produce smaller offspring. In all regards, the early life history differences parallel those of guppies. Their results for senescence were different than what we found for guppies, but are consistent with the more conventional pattern of results seen in other organisms and with Medawar's and Williams's predictions. The rapidly developing *D. pulex* begin reproductive senescence (reduction in the rate of offspring production) at an earlier age, have an earlier acceleration in adult mortality rate and have shorter lifespans than *D. pulicaria*.

Gary Wellborn compared the mortality patterns and life histories of populations of the amphipod *Hyallela azteca* from lakes versus marshes.[34] Lake populations co-occur with fish predators, like sunfish, that prey preferentially on large, adult-size classes of amphipods. The marshes are fish free. The main predator is instead dragonfly larvae, which prey preferentially on small, immature-size classes of amphipods. Wellborn evaluated the life histories of the amphipods both in wild-caught individuals and in the lab-reared grandchildren. The differences in life histories in lakes versus marshes precisely parallel those seen in guppies from high versus low predation environments. The marsh populations are older and larger at maturity and invest less in reproduction than the lake populations. They also produce fewer, larger offspring than their lake-dwelling counterparts.

Many investigators have used the fruit fly, *Drosophila melanogaster*, to perform experimental studies of life history evolution. A number of authors have executed experiments in which they manipulated the risk of mortality, then evaluated how the life history evolved in response to the manipulation. The general result of these experiments is consistent with what we see in guppies. One of the earliest such efforts was executed by Michael Rose, who manipulated early life reproductive success.[35] Selection in favour of high reproductive success early versus late in life produced populations that had higher fecundity early in life than those that were selected for late-life reproductive success, as we find in guppies from high predation environments. The early lines also evolved a more rapid decline in fecundity and more rapid acceleration in mortality rate with age than the late lines, with the consequence that they had much shorter life spans. In this regard, they are similar to *Daphnia*, but different from guppies.

S. C. Stearns et al. performed a similar experimental study of life history evolution in *Drosophila melanogaster*, but instead imposed a form of selection that bears more similarity to what guppies experience in high versus low predation environments.[36] They manipulated the risk of mortality in adults while at the same time rigorously controlling population density. Their high mortality rate lines attained maturity at an earlier age, produced more eggs early in life, but also evolved shorter life spans. However, they also found that their high mortality rate line had higher fecundity than the low mortality rate line throughout their lives. In this regard, their results are similar to guppies.

The *Drosophila* experimental evolution literature has also shown that ecological context can interact with risk of mortality in shaping how the life history evolves.

L. S. Luckinbill and M. J. Clare found that selection on late-life reproductive success causes the evolution of later senescence if larval density is high, but has no effect on the evolution of senescence if larval density is low.[37] Stearns et al. succeeded in selecting for the evolution of the early life history and senescence in response to the manipulation of adult mortality risk, but only after increasing larval density and decreasing food supply.[38] These results, and those for guppies, show that there is an as yet not well characterized interaction between mortality selection, population density and resource availability in shaping the evolution of the life history.

A fourth example involves the activities of humans. Humans exploit natural populations of fish as a source of food. They have proven to be very efficient predators and often prey selectively on the largest prey. The well-developed discipline of fisheries management was built on the implicit assumption that the exploited populations of fish do not evolve, perhaps because it was assumed that evolution is too slow and/or because we did not imagine that humans could have a sufficiently large impact to cause exploited populations to evolve. There is now evidence that many exploited populations of fish have experienced a similar pattern of evolution of smaller size and earlier age at maturity, as seen in guppies.[39]

Interspecific Comparisons

Some investigators have mined life history data on a diversity of species from the literature, and then compiled them into a single analysis that probes for statistical relationships among different components of the life history, or between the life history and features of the environment. The most recent versions of these analyses incorporate the evolutionary relationships among the species, as inferred from DNA-based phylogenies, into the analysis. Doing so makes it possible to infer statistical relationships among different features of the life history without confounding them with the relatedness of the species used in comparisons. One pattern that has emerged in many studies is that the life histories of groups of organisms like mammals or birds array along what has been referred to as a fast–slow continuum. At one end of the gradient are species that mature at an early age, devote abundant resources to reproduction, produce many offspring and are short-lived. At the other end are organisms that are old at maturity, devote less to reproduction, often by producing few, well-provisioned offspring and are longer-lived. These studies show that the life history of any given species is not a random aggregation of different components. Life histories instead evolve in an organized fashion, meaning that the way any one feature of the life history evolves is well correlated with the way other components of the life history evolve. These correlations among different features of the life history are often consistent with the idea of tradeoffs.

In one recent study, G. Peron et al. analysed the association between senescence and the early life history.[40] They compiled data from the published literature on eighty-one free-ranging populations of seventy-two species of birds and mammals. They evaluated the association between events early in the life history and the age at the onset of senescence, defined as the age when there was an acceleration in mortality rate. They found that the early life history predicted ⅔ of the variation in the age of onset of senescence and, specifically, that higher juvenile mortality, an earlier age at first reproduction and the production of more offspring early in life combined to predicted an earlier onset of senescence. This pattern of association is as predicted by Williams[41] and is consistent with antagonistic pleiotropy, or a more general trade off between investment in reproduction early in life with the future potential to reproduce.

Where Else might we Find such Life History Evolution?

A different way of asking about the potential generality of what we have found in guppies is to ask whether or not we often see natural variation in the features of the environment that might select for these same patterns of life history evolution. We do. The field of ecology is replete with studies of the way an individual species can have a substantial impact on the structure of its ecosystem. When this occurs, the presence or absence of the species can cause dramatic changes in the relative abundance of other, interacting species and the resources that are available to them. Ecologists have generated hundreds of descriptions of such numerical responses, but there has been little consideration of their evolutionary consequences.

One such ecological phenomenon is the 'trophic cascade', or the way we can see remarkable changes in the relative abundance of plants, primary consumers (herbivores) and secondary consumers (omnivores and predators) in response to the manipulation of any one of the community members.[42] The existence of such cascades means that differences among localities in the presence or absence of a given species will generate changes in the abundance of other organisms throughout the cascade and hence in their risks of mortality and in the resources that are available to them.

A related phenomenon is the 'keystone species', or a single species whose presence or absence causes significant changes in the abundance of many other members of the community. The classic example is Paine's studies of the effects of the starfish *Pisaster* on the structure of intertidal communities.[43] When *Pisaster* is absent, the community can become a dense monoculture of mussels. When present, there is instead a diverse community. Keystone species have now been described for a diversity of species that occupy different positions in the trophic hierarchy and for a diversity of communities.[44]

A third phenomenon is the ecosystem engineer, or an organism which, by its activities, can alter the landscape in a way that changes ecosystem processes, species composition, and the population densities of individual species. Beavers are one example, because they create new types of habitat.[45] *Prochilodus mariae* is a detritivorous fish abundant in some streams in the Andes Mountains. As part of its foraging activities it clears rock surfaces of sediment and hence exposes them to a diversity of organisms that settle on hard substrates, which in turn alters the abundance of those organisms that feed upon them.[46]

A fourth venue in which we can see strong ecological interactions is where humans have eliminated predators from natural communities. The extirpation of wolves from Yellowstone caused a cascade of changes to the ecosystem.[47] First, elk populations expanded. Their increased numbers and changes in foraging altered the structure of the riparian plant communities which in turn changed the structure of the bird and insect communities.[48] Conversely, when wolves were present their activities as predators changed the structure of their ecosystem. Wolves held the populations of large herbivores in check, which in turn caused a cascading effect on the structure of the rest of the ecosystem. Similar cascading effects of the elimination of predators on natural ecosystems were recently summarized by J. A. Estes et al.[49]

2 IF WE HAVE SEX, DO WE HAVE TO DIE?

Ronald de Sousa

Though you care about the fate of the Earth and the future of humanity, you may not feel that the perpetuation of the species in any way compensates for the annihilation of your own individual consciousness. And as for caring about your progeny, the fact that they will survive you makes it all the more difficult to bear the prospect of your life's end, on the assumption that you will never know their future and that they, for all your faults, may well miss you. In short, like Woody Allen, you aspire to immortality not by your works, or your progeny, or the daisies you will push up from below, but 'by not dying'.

If those are your sentiments, your attitudes are likely to have been formed by a culture steeped in theistic assumptions. For if you were a Daoist, in the tradition of Lao Zi and Zhuang Zi, or even an atheist mystic in the mould of Leonard Angel, you might feel happy with the thought that Life will go on after your individual life has ceased.[1] You might feel that the merging of your individual self in the totality of the universe renders trivial your preoccupation with individual existence. You would have no use for the elaborate intellectual edifices that attempt to make sense of individual survival.

The construction of stories about ways in which we might transcend our own death represents an impressive intellectual achievement. To imagine that I can survive my death, I must first have become aware that I will die. At some time in the past of our species, that discovery made us different from all other animals. Recent reports of what appeared to be mourning rituals among elephants or cetaceans may well be correct, but they do not prove that our mammalian cousins have acquired the awareness of their own future death. They may only know the loss of another. Having discovered death, religions add the further achievement of simple denial.[2] This yields two great benefits for religion. First, it affords an ambivalent comfort for the fear of death. Eternity is repossessed. Second, it provides an impressive tool of control, in the implied promises and threats of post mortem rewards and punishments.

Chief among the aspects of human behaviour targeted by these threats and promises is sex. Sexuality affords innumerable occasions for *mortal* sin: Aquinas, for example, lists no fewer than six distinct species of the sin of lust. Of these, the 'unnatural' sin of masturbation is ranked the worst, as it is by Kant, on the grounds that it represents a deviation from the essential procreative function of sex. For those thinkers, it seems, sex is acceptable only when it represents a defiance of death.[3]

It is not only in religion but everywhere in legend and myth that sex and death seem to be indissolubly linked. Perhaps Tristan and Isolde have to die merely because if they live to a domestic old age it would make a less exciting story; but we may surmise that some deep association in our own minds dictates that aesthetic demand.

For a number of organisms, sex, for the male, involves possible or certain death. As Marlene Zuk has observed, for example, the peculiar heart-shaped mating position of certain species of dragonfly has led evolutionary biologists to speculate that the male needed to keep as great a distance between his body and his mate's as was compatible with insemination to avoid sudden death.[4] And Jennifer Ackerman has written that given 'the "predatory proclivity" of some female dragonflies to banquet off their partners [, m]ales may have adopted the tandem position to protect themselves from becoming their lovers' prey'.[5] Appealing though this somewhat ghoulish topic might seem, what I have to say will shed but limited light on it. What will concern me here is the possibility that the logic of natural selection affords a specific link between sex and death. If such a link exists, it might in some unconscious way underlie the insistence with which theistic religion has consistently striven to control sexuality. It turns out, surprisingly, that while biology is unequivocally on the side of Daoism concerning the issue of the survival of individual consciousness, it can provide solid reason to associate sex with the necessity of individual death. Sex and death are bound up in a common destiny very different from that which, for believers, is symbolized by the serpent in the Garden of Eden, but no less indissoluble and more interesting.

The Common Roots of Science and Religion

Amid the noisy squabbles that surround the relationship between science and religion, it is easy to forget that both have a common source in a very specific capacity of the human mind: a disposition to invent relatively simpler but unobservable entities and processes to explain the complicated world of experience. If we gave up all talk of what cannot be seen, heard or touched, we would be rid of spirits, spells, divinities and acts of faith. By the same token, however, we would have to abjure electrons, genes, quarks and all other entities of which science treats but of which we directly perceive only effects.

For the religious mind, the unobservable entities posited are *agents*, who wish us well or ill. Mostly ill, in fact: the most plausible explanations our ancestors devised of the random disasters of life involved divinities whose ferocity could be mitigated only at the cost of bloodthirsty sacrifices. But by what psychoanalysts call a 'reaction formation', divine cruelty has, in some religious sects, been transmuted into perfect benevolence. This would seem to confirm a central claim made by Pascal Boyer about the nature of religious faith, namely that religious ideas are most likely to take root in the minds of human beings when the familiarity of humanlike agency is spiced up by implausibility. Unlikely stories are more memorable.[6]

The inversion of divine motivation from evil caprice to perfect benevolence retains the key teleological feature that separates religion from science: the positing of unseen agents. Modern science begins when it gives up the hypothesis that hidden causes are agents. Some systems of thought, notably Aristotle's, appear to be transitional in retaining the idea of teleology as part of the natural world without requiring to be backed by spiritual or immaterial agents. But physics and chemistry take their modern form as fundamental sciences by excluding teleology altogether. That is more difficult to do in biology. This is a point to which I shall shortly return.

A second requirement for dislodging science from religion was to forge links between imagination and the empirical data that science aims to explain. These links required to be bound by *reason*, *reasoning* and the *reasonable*. These notions do not quite coincide, and they are difficult to codify; for that reason they are not all that difficult to counterfeit. Indeed, some would claim that the perennial philosophical industry illustrates just such counterfeiting. Despite its vaunted allegiance to reason and reasoning, philosophy has all too often presented the unreasonable spectacle of an intellectual fifth wheel turning idly down the centuries.

However that may be, disciplined reasoning has allowed science to uncover large number of hidden mechanisms in nature. Reasoning is seldom compelling: alternative interpretations are always possible. In 1774, Antoine de Lavoisier rebelled against the wanton tolerance of arbitrary hypotheses wheeled in to save standing beliefs without regard to plausibility or consistency. Calx formed by heating mercury got heavier rather than lighter from phlogiston's supposed release, so it was posited that sometimes (though not always) phlogiston has negative mass. Further experiments convinced Lavoisier that the facts supported a simpler hypothesis: that heat adds oxygen rather than releasing phlogiston. It now seems obvious that this was the better theory. But it shows that auxiliary hypotheses can never be ruled out by logic alone, and that a theory can generate fruitful puzzles. In this case, the bizarre contortions generated by the phlogiston theory set up the challenge to which Lavoisier responded. Similarly, theologians confronting the apparent conflict between faith in a benevolent and omnipotent God and the existence of natural catastrophes can discard the divinity. However, they more commonly fall back on a quiver-full of alternative hypotheses: that God has sent suffering to try us; or better, that the designs of the Almighty are beyond us. (Which conveniently absolves us from the need to reconcile contradictions.) In science, as in theology, one can always get around an unfavourable result. The difference is that in science a contradiction is a problem, whereas in religion it is a solution, renamed Sacred Mystery.

Among such religious Mysteries are the Mystery of the Three-in-One God and the 'Problem of Evil'. These illustrate one more feature common to both religion and science: for both, solutions generate their own problems. You first have to come up with the idea of the Trinity before you can worry about making sense of a God that is both one and three. And without the postulate of theism, there is no problem of evil. Nature of itself embodies no values. It is unconcerned with the happiness or the suffering of the creatures that for the last billion years have preyed on one another. One may well find this fact distressing, but only on the assumption that the universe is ruled by a specific sort of God does it constitute an intellectual *problem*.

For science, sexuality, death and the relation between them are comparable to the Trinity or Evil for theology: they are problems manufactured in-house. Like suffering, sex and death cause a good deal of trouble in everyday life. They become intellectually problematic only on the basis of certain assumptions made by biology itself.

What are these self-made problems? Once one gave up pre-formation, which saw in the spermatozoon a miniature but fully formed adult, it has been accepted that sexual reproduction mixes up genes at every generation. Every gamete, or sex cell ready for merging, results from meiosis, the process that divides the genetic heritage of each parent into two. The male and female gametes resulting from meiosis merge to form a zygote, a single cell in which the two halves of the male and female legacy are united, and which divide, first by simple cleavage and later by mitosis, to form all the cells of an absolutely novel individual body. In this way, sexual reproduction requires that a successful model, constituted by any organism that has proved viable, will be abandoned, never to be copied again. At every generation, regardless of the qualities exhibited by the previous unique prototype, the mould is broken. This seems wasteful, or at least extraordinarily risky: two perfectly successful models, instead of being reproduced, are forced to make way for an untested model.

Sexual reproduction also exacts what the evolutionary biologist John Maynard Smith described as a twofold cost.[7] Various species of geckos and lizards in the deserts of the American South West, among other species, seem to have taken advantage of this fact.[8] They have won out over their sexually reproducing cousins by resorting entirely to parthenogenesis. By this expedient they need only half the resources their cousins require, since the latter must feed two parents instead of one for the same number of offspring. So why do we not do the same? What are the compensating advantages of sex?

Such are the problems that evolutionary biology sets for itself about sex. As for death, once one ceases to deny its existence, one can still ask: what is death for? Here again, it is not a matter of deploring the practical inconveniences involved. Rather, reflecting on the ironically reference-free phrase, 'survival of the fittest', the question is why in fact, absolutely no sexually reproducing organism ever does survive. Would it not have been more economical for natural selection to have indeed selected the fittest for *perpetual* survival?

Two Kinds of Possibility

The question rests on two presuppositions that may be debatable, but which I take to be both correct and important. One is that it is appropriate to speak of biology in terms of economics; the second is that we can make sense of natural, biological teleology in the absence of design. The justification of the first assumption is that economics works better for biology than for economics. The reason is simple: economic 'laws' work only if we make auxiliary psychological assumptions concerning motivation. These often prove false. In biology, by contrast, the relevant gains, losses and probabilities can be directly computed in terms of the number of genes surviving later generations, and the objective probability conferred on their survival by their capacities in ambient conditions. In order to understand how natural selection drives evolution, there is no need to speculate on the motives, emotions, conventional constraints on greed, or anything else that might inhibit rational maximizing strategies.

The second issue is trickier. The question, why? often confounds causality and teleology. For all metazoan organisms – multicellular organisms with functionally differentiated cells – decline and death are part of natural life. For Aristotle, any complete explanation worthy of the name must include a teleological component; every creature, as well as every organ, has its own proper job to do. Among all the different things that are possible, there is a special class, labelled *potentialities*, which alone are *supposed to* become actualized. It is possible for the acorn to rot; it is also possible for it to grow into an oak tree. Only the latter constitutes its potentiality, what is *supposed* to happen.

But how can we tell which, among all possible events, is the one that is supposed to take place? Aristotle thought the answer could be derived from simple observation. On the assumption that species are more or less immutable, what is statistically normal is also *normative*. In this way, what happens 'always or for the most part' is a sufficient guide to what nature 'intends'.[9]

Modern science has abjured the teleological perspective inherent in Aristotle's worldview. Yet perhaps Aristotle's demand for teleological explanation expresses a basic intellectual craving. Perhaps we still yearn for a narrative that linked sex and death by telling us what they are both *for*. After Darwin, however, no such divining of the inherent intentions of nature makes sense. Nature is mindless. Worse, Aristotle's recipe leads to a particularly paradoxical conclusion in the light of Darwin: if all our ancestors had functioned properly, that is, if all had conformed to what happens 'always, or for the most part', we would still be, like the vast majority of living things, unicellular organisms. Every ancestor whose difference brought us a little closer to being human was necessarily, in Aristotelian terms, a monster. None could be reckoned to be what was 'supposed to happen'. And of our current deviancies and perversions, none can say which will be hailed in retrospect as progress.

It is therefore quite true that in a certain sense Darwinism has banished teleology. Evolution is a random walk through the space of possible forms. Nevertheless, in what may well be that rare phenomenon, a genuine philosophical advance, analytic philosophy of biology has constructed, over the past fifty years, a purely naturalistic version of the teleological notion of 'objective function'. In this reconstruction, some actual or possible effects are distinguished, in a manner that recalls in Aristotle's view, as those that are supposed to happen, on the basis of past facts. In spite of that similarity, there are three crucial differences – amounting to an abyss – between Aristotle's conception and the modern one. First, those effects that we can label as functions are not part of an essence or nature that is manifested always or for the most part. As Ruth Millikan has pointed out, we do not hesitate to assume that the function of a spermatozoon is to fertilize an ovum, though only one in a billion or more ever does so.[10] According to the *aetiological* conception of teleology – so called because it is based entirely on causation – a function is simply an effect that has contributed to the differentially successful reproduction of similar ancestral organisms. This notion of function is entirely objective: it does not depend on our interests or on our perspective. That is not to say that it is easy to detect in all cases, since it appeals to differential advantages in ancestral populations which cannot be verified directly. But it is a matter of fact even when it is unverifiable that the capacity of an organ to produce certain specific effects did or did not give the organisms that contained it an adaptive advantage.

Take the heart, for example. We can observe two effects of its activity. One is the circulation of the blood; the other is the production of rhythmic sounds audible through a stethoscope and useful, as it happens, for various diagnostic purposes. Which of these constitutes the function of the heart? According to the aetiological criterion, the use we make of heart sounds is probably irrelevant. The reason is that only the former effect presumably afforded an adaptive advantage to ancestors of organisms currently equipped with a heart. That does not necessarily exclude the possibility that heart sounds also had certain advantages for organisms equipped with hearts. However, whatever the actual facts of the matter may be, what it means to say that the heart's primary function is to circulate the blood, is that blood circulation was more important than rhythmic sounds to the successful dissemination of genes for having a heart. The aetiological conception of function has given rise to a number of criticisms and refinements. In the main, however, it has remained unshaken, and it is not necessary here to rehearse the debates and intellectual meanderings to which it has given rise.[11] The simplified account I have given suffices to show how this conception explicates a viable notion of teleology for biology.

A second difference between Aristotle's story and the one I have just sketched is this: for Aristotle, the fulfillment of the teleological goal – the actualization of a natural potentiality – automatically embodies a positive value. The same is not true for the objective functions I have characterized. Sometimes, perhaps often, the exercise of a natural function is a Good Thing. However, many natural dispositions that are plausibly regarded as functions are obviously to be deplored. Predispositions to xenophobia, for example, or to rape, may well in the past have contributed to the evolutionary success of our ancestors. But that is no reason to approve of them now. The taste for fatty salty and sweet foods so effectively exploited by fast food chains is certainly an adaptive trait that once functioned to steer us to foods that were both necessary to our thriving and hard to find. Again this is not a tendency which currently serves us well. Its function is objective, but not thereby to be valued positively.

A third important difference is the following: for Aristotle, teleology applied not only to organs but also to whole organisms. By contrast, the logic implicit in the idea of an objective function requires that the beneficiary of that function be external to the entity that executes it. So the idea that an individual organism as such has a function is problematic. It makes sense only insofar as we consider the role of an individual in relation to a larger entity such as a group, a society, a corporation – in short, any larger unit in which the individual functions like an organ in an organism. Etymologically, an organ is a tool. I might be serving as a tool for the purposes of some entity of which I am not even aware. But in myself, as an individual considered as such, it makes no sense to say that I have a natural function.

Perhaps, however, individual organisms should not be regarded simply 'as such'. Perhaps there is a way of regarding metazoan individuals that allows them to play the role of 'organs' in the context of a larger unity. This way of thinking is familiar to students of bees and other social insects who have followed the pioneering work of Edward O. Wilson: in relation to the hive, each individual can be regarded as an organ with a function.[12] If human society is viewed in that light, one might think of birth and death as functions fulfilled by individuals in the service of the collective entity.

From the point of view of a modern individualistic ideology, that perspective may seem alarmingly reminiscent of certain collectivist ideologies that have been all too happy to sacrifice individual humans to a supposed social whole. But we do not need to think of the entity we serve as a larger, societal entity. One can instead think of it as consisting in smaller units. Genes might be just the thing: they are the only entities that actually survive, transcending not only the puny life of individuals but even the species. It is not the fittest individuals that survive, but the fittest genes. As such, they influence the course of individual lives, in both morphology and behaviour. We can therefore regard the propagation of genes as constituting the goal in relation to which the individual human being is a tool fulfilling a function.

Richard Dawkins has spoken metaphorically of the individual organism as a vehicle used for their own purposes by the genes that inhabit it temporarily, and programmed to facilitate their perpetuation. By 'genes', I mean, with Dawkins, not only specific sequences of DNA mapped by the cartographers of the genome and entrusted with the programming of protein manufacture. The term must be understood as referring to any informational complexes that are transmitted from one generation to the next.[13] As Susan Oyama has urged and most now agree, these may include more than the codes embodied in DNA sequences.[14] Under the general designation of 'epigenetics', some biologists have reported evidence for mechanisms of transmission may work directly through the cell structure independently of DNA.[15] However, regardless of what the mechanisms of heredity consist in, the informational entities that get transmitted are privileged in relation to the perishable individual organism that is constructed on the basis of the various heritable mechanisms, as well as of the interaction between them and a complex environment.

So far, nothing has been said to imply the necessity either of death or of sexuality. If the 'goal' of my genes is to survive, would not the best means to this be to make the vehicle that carries them last forever? What then is the role played by death and sex in this vision of things?

There is No Law of Biology that Decrees that All Organisms Must Die

Let me begin by disposing of two pseudo solutions. One finds too much teleology in death, while the other ascribes too little.

Some, including the French Jesuit evolutionist Teilhard de Chardin, have seen death as facilitating the great epic of evolution construed as a march towards a clear destination, culminating in a creature made in the image of God.[16] In this conception of evolution, of which many versions have recently taken to huddling under the capacious tent of 'intelligent design',[17] natural selection is responsible for the easy bits in the uneven progress towards complexity and perfection. The divine intelligence which Darwin did without reappears either to steer the whole enterprise in the right general direction, or as the 'God of the gaps', to produce the occasional miraculous saltation to some new plateau of what Michael Behe has called 'irreducible complexity'.[18] On this view, death is required in order to make room for new prototypes. Even a sophisticated evolutionist can sometimes be caught saying something that sounds rather like that. Jacques Ruffié, for example, in a book entitled *Sex and Death*, no less, writes that 'sexual reproduction ceaselessly creates new types ... But these are able to disseminate new combinations and forms only if the old ones make room for them.'[19] There is nothing wrong with this formula, so long as it is interpreted as a mere observation of fact. If the earth's every niche were fully occupied by a fixed number of immortal individuals, any experimental novelties would be crowded out. But if Ruffié's observation is intended as more than a platitude, it seems to commit two rather common fallacies. The first consists in presupposing that evolution as a whole has a general direction towards a predetermined goal. The second is to assume that complex forms are inherently superior. This is both subjective and self-serving.[20] The only objective criterion of success is persistence through the aeons. In light of that criterion, cockroaches and bacteria have been by far the most successful forms of life.

The most attractive alternative in the face of these disappointing attempts to wring a purpose to life out of the facts of evolution is simply to fall back on the null hypothesis, which refers to situations in which a phenomenon of interest is merely the product of chance rather than the effect of some specific cause. On the null hypothesis, there is literally nothing to explain. Newton's first law provides a handy illustration. Ever since the Ancients, philosophers had worried about the question: what keeps the arrow in-flight? Newton's first law simply rejects the question, and replaces it with its inverse: what is it that *impedes* the arrow from continuing in uniform flight? The first law states that a body in a state of uniform motion in relation to an inertial system will continue in that state unless acted upon by a force. The null hypothesis about death, then, is that death has no function whatsoever in the objective sense of the term I have sketched. Rather, it is merely the consequence of an absence of special causes or conditions that would bring about the long-term persistence of individuals. A number of considerations lend plausibility to the null hypothesis

It is known that – with some specific exceptions which will concern us in a moment – cells set to divide freely in vitro will continue to do so only about fifty times. Thereafter, the copying process peters out, as if copies had degraded to the point where they are incapable of further reproduction. If we remember that any process of copying, however careful, involves a certain probability of introducing copying errors, this suggests what might be called the xerox-copy model. Faithful reproduction to any given number of copying generations can be secured by a suitable increase in the redundancy of the information to be copied. (That is why, when telegraphing a number, it is advisable to send it in letters as well as in figures). But the introduction of redundancy is not cost-free. In the case of sexually reproducing organisms, natural selection is likely to prefer those organisms that have enough redundancy in their cell-reproducing mechanisms to endure long enough to produce a new generation. However, once the organism has actually reached sexual maturity and produced offspring, the genes it has passed on are more efficiently conserved and reproduced by those offspring than by the continuation of the parents' own reproductive activity. The reason is that the latter option would demand a higher level of redundancy in the parents' own somatic cells. This suggests the following biological surmise: sex cells provide the intact archival copy that is referred to when new sets of copies are to be made, and death is a side-effect of the limited redundancy of genetic information encoded in each somatic cell. At every generation, copies must be made from the original – the sex cells – because the copies already made of somatic cells in this generation are no longer capable of producing more. Death is in no way an adaptation; rather it is merely the absence of any 'adaptation' that would entail perpetual life. The xerox-copy model can be regarded as the null hypothesis regarding the existence of death.

Biological reproduction reaches astonishing degrees of accuracy. Mark Ridley has estimated that in each transmission of a human genome, for example, one can expect about 200 errors.[21] The human genome comprises some 30,000 genes, made up of about 3.08×10^9 base pairs. 200 errors therefore represent two out of 6.16×10^7, or an error rate of 0.0000000325 per cent.[22] This is already a spectacularly low error rate, but it applies to sex cells, not to somatic cells. We shall see in a moment what the significance of the distinction might be. For the moment, suffice to say that if we remember that somatic cells need to reproduce themselves by mitosis on a very frequent basis, even a very low error rate will result eventually in deleterious changes in cells' properties. Sooner or later, these errors will prove lethal. Introducing further protection against error, in the form of redundancy or in any other way nature might devise, is necessarily going to be costly. And the inexorably economic point of view of natural selection will impose an equilibrium between the cost of exactitude and the benefits that might result from it.

What benefits might these be? In a sexually reproducing species, it is obvious that the utility of the mature individual to the dissemination of its genes lasts only as long as it is capable of procreating. Only genes are literally reproduced. The individual that constitutes the vehicle, after completing its mission of gene transmission, is expendable. It is therefore to be expected, without appealing to any hypothesis about what might be the inherent benefits of death, that the level of redundancy operative in any given species will protect somatic cells from degeneracy just long enough to ensure the transmission of the sex sells containing the faithfully copied genes. Somatic cells last only so long and they are regularly replaced by fresh cells, arising either from division or from stem cells capable of adopting any functional role. It is now known that the limitation to fifty divisions depends on the gradual shortening of telomeres, the sort of expendable plug at the end of a chromosome that provide a necessary buffer for complete copying of what lies in between.[23] That number of divisions takes about 120 years. This suggests that natural selection has equipped the human body with just enough endurance to last about two or three times the maximal duration of the reproductive cycle. Like any good engineer, it would then have provided enough of a margin to guarantee the performance of the essential reproductive task. This will satisfy both the demands of quality control and those of planned obsolescence, designed to avoid wasting resources on the maintenance of the tools of gene transmission beyond their useful life. Scientists may well soon discover a way to protect the telomeres from erosion, thus transcending the limit in question; but as things stand in nature, we can expect a normal human organism to break down after its cells are no longer able to reproduce normally. Such an explanation, which amounts to the rejection of a demand for explanation, is both plausible and economical.

Nevertheless, there are reasons to believe that there might, after all, exist a tighter link between the sexual reproduction and the regular death of all individuals. Two factors are particularly worthy of attention.

The first is the phenomenon of *apoptosis* or programmed cell suicide. When a cell receives an appropriate signal, or rather, which comes to the same thing, when it ceases to receive a signal that serves to avert apoptosis, the process of self-destruction automatically begins. Jean Claude Ameisen has described this process as essential to the construction of bodily structures, particularly very complex ones such as the brain and the immune system. Although it might seem paradoxical, suicide at the level of the cell guarantees the integrity of the body up until the moment that the body as a whole is sacrificed in death.[24] The very existence of apoptosis casts doubt on the null hypothesis. The null hypothesis explains the fact that cells cease to reproduce themselves with sufficient accuracy as a simple consequence of the limited protection afforded by redundant informational resources: after a while, that protection fails. At first sight, that seems incompatible with the existence of a specific mechanism for cellular auto-destruction.

Actually this is not enough to establish that the null hypothesis is false. Death of the organism might still be ultimately due to the degeneration of the information required to keep its cells alive. For if apoptosis is essential to the construction and maintenance of the body, it may itself be one of the processes that ceases to function properly after a certain amount of wear and tear.

An additional piece of corroborating evidence is therefore desirable. And it can be found in the existence of three exceptions to the general rule that says that cells can only divide successfully a fixed number of times.

The first exception concerns bacteria. For these, unless their reproduction by division counts as death, there is no natural death. Cancer cells constitute the second exception. They behave as if they had become deaf to the regulatory signals that rule the community of cells by controlling mitosis or mandating apoptosis.

Cancer cells seem to have unilaterally abrogated the contract that binds every specialized cell to the organism of which it is a part, and to which it owes allegiance, insofar as it cannot survive alone. By killing off the organism, cancer cells ultimately destroy themselves, so their rebellion is self-defeating. Unless, of course, their potentially immortal life is sustained by being used in research, like those of the black cancer patient, Henrietta Lacks, whose story was told by Rebecca Skloot.[25] The third exception, of course, is the lineage of sexual cells from which gametes emerge. In each organism, these cells are kept isolated from others, and passed from one perishable body to another by means of sexual reproduction.

What is demonstrated by the existence of these three sorts of lineages of potentially immortal cells is that natural death is not an absolute rule of nature. It is only among creatures like us, that is, metazoan creatures that reproduce sexually, that the death of individual organisms is an inevitable part of life. Somatic cells are copies which beget more copies of themselves, just as sexual cells do, but the former, once they are specialized for the specific tasks required of the organs they constitute, become more vulnerable to changes that will impair their functioning to the point of uselessness. They will not, however, transmit the defects they have acquired in the course of doing their job. Instead, they will disappear when they have exhausted their capacity to serve their function. The sexual cell lineage, by contrast, will devote its entire efforts to ensuring the accuracy of the copies they make of themselves, and in so doing they will avoid the wear and tear to which somatic cells are subject, and escape death itself.[26]

The Sex and the Death Complex

Evolution did not take place in order to trace a path to the emergence of us humans, or for any other purpose. It is a blind exploration of the space of possibilities. Sexual reproduction, in particular, followed on two random 'discoveries' that together set new constraints on the life of cells at the same time as they greatly expanded the space of possibilities. The two discoveries in question are *conjugation* and *collaboration* between agglomerated cells.

Cell conjugation allows genetic material to flow between two individual unicellular organisms such as bacteria. Conjugation is unrelated to reproduction: after conjugation, there are still just two cells (though one of them might have benefited, and the other might have been harmed, by the transfer of genetic material). One might say that conjugation was the first recreational sex, discovered by bacteria one or two billion years before sexually reproducing species invented sexual reproduction. That invention already brings some of the advantages that biologists have surmised compensate for the drawbacks of sexual reproduction: some cases of conjugation are thought to result in the repair of one cell's genetic material. According to a once heretical but now generally accepted hypothesis first offered by Lyn Margulis, conjugation affords a cell the opportunity of raiding another in the hope of replacing a defective piece of genetic material.[27] Like sexual reproduction, this mechanism is highly risky, for the change in genetic holdings in one or both of the cells in question can result in damage rather than repair. What is certain is that it is likely to produce something new, and thus to promote diversification.

In the small section of the biosphere inhabited by us metazoans, gene exchange has hit on new functions within the framework of cooperating agglomerated cells capable of reproducing sexually. That is what has brought about the imbrication of death and sex. That interdependence stems from a number of constraints inherent in the nature of metazoans, and notably to 'Weismann's barrier' and what Francis Crick baptized the 'Central Dogma' of molecular biology.[28] Let us briefly see how these factors work to link sex and death.

First, a metazoan organisms is made up of about a ten to a hundred trillion cells, differentiated in both morphology and function into a few hundred types that make up the different organs integrated into an individual body. Some hundred billions of those cells are neurons, organized into a brain whose activity produces consciousness, together with such a powerful illusion of unity that many philosophers from Plato to Descartes have insisted that we possess an indivisible and therefore indestructible individual soul, to which we owe our capacity for rational thought and action. Rationality and individuality are properties that are not easy to reconcile, for the latter suggests that we are all different, while the former seems committed to our being all the same. This follows from the presumption that reason is capable of converging on truth and more generally on the uniquely correct solution to any question that admits of one – at least if we can bring ourselves to formulate the questions fairly and confront the answers clearly. Reason, as Heracleitus is said to have remarked, is common to all, yet each one of us thinks it belongs peculiarly to ourselves.[29]

Elliott Sober and David Wilson have made an ironic comment on our sense of uniqueness, suggesting that natural selection has equipped us with a special feeling of 'ipseity' – the sense that I am different from anybody else – precisely in order to compensate for the fact that there is nothing distinctive about me. This view makes good sense within the economic perspective I have been advocating. It is important that each organism look out for its own interests in the medium term. However, if there is nothing objective that distinguishes my own future or my own interests from those of anyone else, nature must instill in us an illusion of difference in order to save us the trouble of calculating what does and what does not pertain to the interests of the long-term continuant that is my body. Sober and Wilson note the irony: 'people use the concept of "I" to formulate the thought that they are unique. Yet part of the reason that people have this concept is that they are not unique.'[30]

Second, both reason and consciousness are frail capacities. They depend entirely on delicate configurations of our trillion neurons. Even those sorts of experiences that have been taken as proof of our separable spiritual essence, such as 'out of body experiences', in which subjects think they are observing their body from an external point of view, have now been provoked to order by the stimulation of the right group of neurons.[31] If unicellular organisms are conscious, they can hope to retain an unlimited memory of their lives before previous fissions. But sexual reproduction guarantees that individual consciousness will not survive individual death, since the neurons that embody our mental states, including our memories, once dispersed, could never be reassembled to constitute the same brain state again. So if it is individual survival you want, you are out of luck. Better fall back on the Daoist view.

The third piece of supporting evidence relates to Weismann's barrier. As we have seen, by ensuring that information coded in the DNA goes only one way, from DNA to the proteins that are made in somatic cells, and never from protein back to DNA, the mechanism implementing the 'central dogma' sequesters the sex cells. In addition, the sequestration of the sex cells makes it virtually impossible for any Lamarckian process to endow the next generation's genome with changes brought to an organism's somatic cells. From the point of view of the aetiological conception of function sketched above, which alone is compatible with the standpoint of Darwinian evolution, this alliance between somatic death and the preservation of faithfully copied sexual cells can be regarded as the function of Weismann's barrier.

Some qualification is in order: the barrier is not absolute, at least not in the first few divisions of a zygote. As we have seen, the zygote results from the fusion of gametes which themselves issue by meiosis from sex cells segregated from the parental body. The zygote divides repeatedly and ultimately generates cell differentiation and proliferation, giving rise ultimately to the adult organism. But this means that the zygote is the ancestor of both somatic and sex cells. The former will constitute the building blocks of the parts and organs of the living body; the latter will hold on to the genetic material that will ultimately constitute the essential message contained in future gametes. Yet until this division of labour is set up, that is, for the first dozen or so divisions, the zygote gives rise by simple cleavage to cells that are as yet undifferentiated, and hence 'pluripotent', to every kind of cell.[32] A change brought to the DNA of these early cells could be inherited by sex cells and find itself in the following generation. However, this window quickly closes. Soon the destiny of somatic cells will have been sealed, and that destiny, our destiny, in fact, will be that of a cul-de-sac of evolution.

Sexuality therefore condemns us to die, but it also ensures that we will never meet another individual just like ourselves unless we have a 'real' or monozygotic twin. The generality of this fact may explain both the attraction and the occasional horror that is inspired in some by the idea of cloning. To copy an individual human being, many people feel, is to do something that is against nature in a particularly creepy way, prompting politicians to ban the practice pre-emptively even before it has become practicable. Who knows what unicellular politicians might have said, one or two billion years ago, when the first bacterium 'decided' to 'experiment' with sexual reproduction?

I mentioned above two of the notorious drawbacks of sexual reproduction: the twofold cost entailed by the requirement to keep males around and the risks inherent in meiosis. Many hypotheses have been suggested to explain the countervailing advantages of sex. One of them, it will be recalled, is that sex contributes to the diversity of genomes, which may become an advantage under changing conditions, in that it increases the chances that descendants of any particular organism will include some that are pre-adapted to different conditions. But there is something glib about this formula. It leaves unspecified the precise identity of the beneficiaries in question. If this is an advantage, who exactly is it an advantage for? Remember that whoever survives, it will not be *me*. I will not be the one to benefit from the genetic diversity to which my sexual union with another individual might contribute. At this point, one generally switches from talk of the individual to speaking of the species. Yet what motive do I have, as an individual endowed with an individual consciousness, to identify with the interests of a *species*, even if it is my species? Why should I care? I can empathize with another consciousness; but I cannot literally merge with it. So it is not literally in my interests that the species, or my descendants, should survive. Conversely, why should I exclude from my concern other real or imagined consciousnesses with which my own might find itself in harmony? Why should I not find a kind of mystical comfort in what links me to other inhabitants of the biosphere, even if they are unlike me and are not members of my own species?

Just such an attitude is to be found, expressed with poetic panache, by the great biologist David Hamilton, who evoked in his testament the idea of a survival devoid of any individual consciousness. Instead, he relished the thought of being consumed by and scattered into a thousand brilliantly coloured Coprophanaeus beetles:

> They will enter, will bury, will live on my flesh; and in the shape of their children and mine, I will escape death. No worm for me or sordid fly, I will buzz in the dusk like a huge bumblebee. I will be many, buzz even as a swarm of motorbikes, be borne, body by flying body, out into the Brazilian wilderness beneath the stars, lofted under those beautiful and unfused elytra which we will all hold over our backs. So finally I too will shine like a violet ground beetle under a stone.[33]

As a matter of psychological fact, most of us feel more closely concerned by the fate of our own descendants than by that of the Coprophanaeus beetle of Brazil; but that is just the effect of a subtle manipulation of our minds by our genes. In order to maximize their chances of being perpetuated, my genes have programmed me to forget that my children are not identical to me. Strictly speaking, the feeling that my descendants' fate concerns me is an illusion. My genes are not me; their goals (metaphorical though they are) are not my goals. I can choose to endorse those goals, or struggle to resist them. The very existence of that choice implies that if I feel concerned, then I am concerned. In matters of emotion, bootstrapping can work to establish that the objects of my concerns are real merely in virtue of my feeling that they are. But the fact that the fate of my descendants seems so evidently relevant to my concerns is merely the psychological manifestation of that destiny that links the four aspects of metazoan life and death together in the way I have sketched in this essay.

Conclusion

To begin with, purely chance encounters resulted first in the gene exchanges of conjugation, and later in the process that gradually established functional differentiation in agglomerated cells. In order to protect themselves against an accumulation of copying errors in the lineage that included the cells that would give rise to gametes, the sex cells segregated themselves from the somatic cells, which became doomed to die as soon as they had accomplished their mission, thus entailing the necessary death of the individual.[34] The fact that we are individuals and the fact that we must die are aspects of the very same reality.

Is this story just another myth? If it is, how shall we assess it in relation to those that were wont to comfort our ancestors? Some of us, like Hamilton, may find a certain charm in the thought of the unity of life; others may find it repellent. In such aesthetic matters the claims of subjectivity cannot be ruled out of order.

If this story still seems implausible, we should remember that according to Pascal Boyer, as we saw above, implausibility is a characteristic that is liable to improve a myth's chances of survival. However, this particular myth enjoys an additional advantage: that of being veridical. At the very least it has as much claim to being veridical as can accrue to a perspective grounded in scientific fact, accepted with the usual admission of fallibility. The truth has its own poetic allure. Compared to the hopes for survival encouraged by theistic religions, the vision of our destiny afforded by the mutual implication of sex and death seems to me more interestingly surprising, and perhaps ultimately more consoling.

3 THE GENOMIC CONSEQUENCES OF ASEXUAL REPRODUCTION

Ken Kraaijeveld and Jens Bast

Introduction

Sexual reproduction originated about two billion years ago, together with the first multicellular organisms. Sex was maintained over evolutionary time and now the overwhelming majority of eukaryotes reproduce sexually. By contrast, asexual taxa are scattered among the tips of the evolutionary tree of life and are prone to early extinction. This pattern has puzzled evolutionary biologists for decades.[1]

All else being equal, an asexual population consisting only of females should grow in size at twice the rate of a sexual population (the so-called 'twofold cost of sex'; see Figure 3.1). Given this huge advantage, asexuals would be expected to quickly outcompete their sexual counterparts. Indeed, there is evidence that this can happen on a local scale. For example, Lake Naivasha in Africa used to be populated by a sexual population of *Daphnia* waterfleas. In 1920, a single asexual waterflea lineage was introduced into the lake. Sixty years later the asexuals had completely replaced the sexual population. Today, all waterfleas in the lake belong to the same clone and genetic diversity has dropped to zero.[2]

Despite this short-term advantage of rapid population growth, asexual reproduction appears to be an evolutionary dead end. In this chapter, we briefly review the processes that can explain these patterns and present some recent evidence that support these. This is intended as a general discussion only. For a more detailed overview of the various hypotheses see G. Bell and I. Schön et al.[3] We then examine how processes that shape the genome may play out differently in sexuals and asexuals. In particular, we consider two such aspects: transposable elements and epigenetics. Last, we depict some exceptional taxa that appear to have been reproducing without sex for millions of years and that challenge our views on the evolutionary costs and benefits of sex.

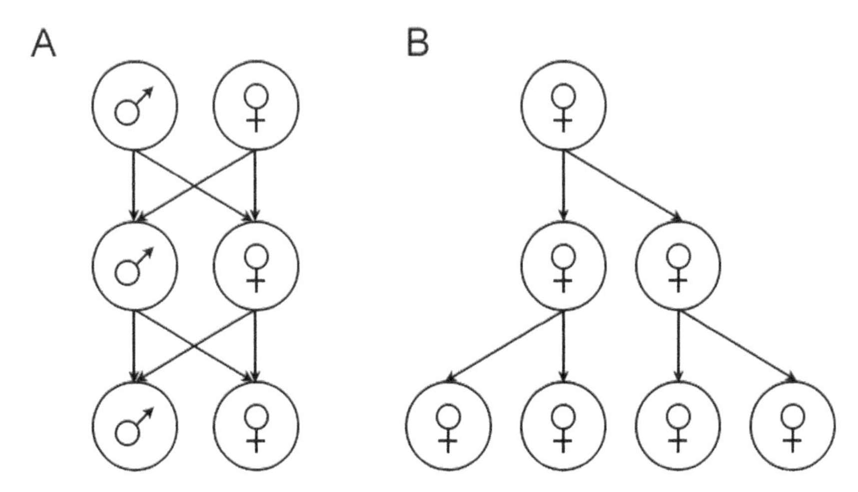

Figure 3.1: The two-fold cost of males in sexual populations: (a) Sexual populations producing 50 per cent males and two offspring per generation will be constant in size; (b) Asexual populations producing two offspring per generation will increase in size exponentially.

Many Ways to be Asexual

In this chapter, 'sex' refers to the mixing of genes through the fusion of gametes produced independently through meiosis within two individuals. There are many mechanisms to reproduce without combining the genomes of two individuals. The simplest way is to bud off a part of the body. A strawberry plant will send out shoots, which develop into new strawberry plants. This process does not include meiosis or genetic recombination. Budding is common in plants, but also occurs in primitive animals such as *Hydra*. At the other end of the scale are selfing hermaphrodites. Like a sexual, a selfing hermaphrodite goes through the process of making sperm and eggs. The difference is that these can fertilize each other, in which case there is no mixing of genes from two individuals. This mechanism is found in many plants and certain animals, including snails.

In between these two extremes is a wide variety of ways of being asexual. We will highlight some examples, but stress that this list is far from exhaustive.[4] Any general theory to account for the disadvantage of asexuality should apply to all these mechanisms. On the other hand, each different mechanism will result in different effects on the genome.

Asexual geckos (*Heteronotia binoei*) in the Australian desert are the descendants of hybrids between two distantly related species.[5] These hybrids could reproduce asexually or could mate with one of the parental species. Mating resulted in three sets of chromosomes instead of the normal two in their offspring. These were mostly asexual, although some would mate with one of the parental species. This produced asexual descendants with four sets of chromosomes. These asexual geckos are very common in the Australian interior. The parental species live around the margins of the continent.

Amazon mollies (*Poecilia formosa*) are freshwater fish found in Mexico and Texas. The Amazons lay eggs that are genetic copies of the mother. However, like the eggs of the sexual ancestors, their development needs to be triggered by sperm cell penetration. This is problematic, because there are no male Amazon mollies to provide sperm cells. Asexual females solve this problem by mating with males of several different species. Male sperm cells help to kick-start the development of the Amazon eggs, but their genes have no part in the resulting offspring.[6]

One astonishing animal group is the parasitoid wasp (e.g. *Leptopilina*). Like related bees and ants, wasps reproduce through haplodiploidy, a peculiar form of sexuality. Females in these insects either fertilize their eggs or not. If fertilized, the egg will develop as a daughter carrying the genetic material of both the mother and the father (as in other sexuals). Eggs that are not fertilized develop as males that carry only a single set of maternally derived chromosomes. This system is sometimes hijacked by selfish, asexuality-inducing bacteria. *Wolbachia* is a bacterium that lives inside the cells of many insects, including many parasitoid wasps. To get from one wasp-generation to the next, *Wolbachia* is completely reliant on egg cells. Sperm cells are too small to harbour the bacteria. That means that being in a male is a dead end for the bacterium. In parasitoid wasps *Wolbachia* assures that it resides in a female in a very effective way by turning unfertilized, haploid eggs into diploid eggs. For example, in the wasp *Leptopilina clavipes* (see Figure 2a), *Wolbachia* interferes with the first cell division of an unfertilized egg in which it lives, so that the chromosomes double as normal, but do not divide over two daughter cells.[7] This results in a single cell with a double chromosome set, which develops as a female.

One of the most bizarre examples of asexuality is found in little fire ants (*Wasmannia auropunctata*) from South America. Genetic fingerprints of many individuals from several colonies revealed two strange things.[8] First, all the queens (the matriarchs of ant societies) were completely homozygous. Their two sets of chromosomes were identical. Like the *Wolbachia*-infected parasitoid wasps, they had to have developed from unfertilized eggs in which the single chromosome set was somehow doubled. What was even stranger was that the males were genetically completely different from their mothers. The explanation, which still awaits confirmation, is that the males develop from fertilized eggs. These contain the genes from both the mother and the father. If something destroys the mother's gene complement, the egg will have only one set and thus develop as a male. Asexual males are extremely rare in the animal world, but here we have an example. They are living side by side asexual females that are their social and sexual partners, but are genetically something completely different.

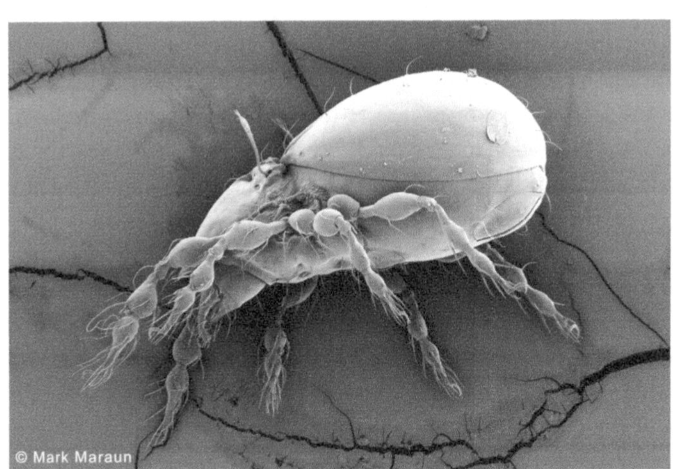

Figure 3.2: (a) top; A female *Leptopilina clavipes* inspecting a fruitfly larva. In some populations of this parasitoid wasp, all individuals are infected with *Wolbachia* bacteria that turn them asexual; (b) bottom; REM picture of *Oppiella nova*, a scandalous long term asexual. The all-female populations existed without sex for millions of years.

These are just a few examples of asexual reproduction. We could easily cite many more. There are organisms such as aphids that reproduce asexually most of the time, but go through periods of sexual reproduction. There are systems in which haploid egg cells turn diploid by fusing with one of the polar bodies. And so on. The point is that asexuality is fairly easy to evolve and has done so regularly and in a variety of ways. Still, all asexuals are descended from sexual ancestors and appear to be short-lived on an evolutionary timescale.

What a Difference Sex Makes

Remember, we are looking for some highly significant advantage of sex that can overcome its twofold cost. There are many different ways to share genetic material between individuals, but for now let us consider sex the way we do it: with separate males and females, each with two copies of every chromosome. Our imaginary organism has only one chromosome. Each egg cell and each sperm cell gets only one of the two copies of this chromosome. A female can thus make two types of egg cells and a male can make two kinds of sperm cells. After mating, there are thus four possible combinations: four genetically different offspring.

Of course we humans do not have one chromosome, we have twenty-three. Each female and each male has forty-six chromosomes to play with. These can be combined into one egg or sperm cell in 2^{23} ways: a female can make more than 8,000,000 different egg cells and a male can make more than 8,000,000 different sperm cells. Combining the two gives no less than 64×10^{12} possible offspring. This means that a couple can make 64×10^{12} children, which would all be genetically different!

And that is not all. During the production of egg cells and sperm cells the two copies of the same chromosome line up and exchange parts (crossing-over). In this way one piece of chromosome copy 1 ends up attached to chromosome copy 2 and vice versa. This introduces variation even within chromosomes. A female has thus many more than forty-six types of chromosomes to play with. And that makes the number of possible offspring basically infinite.

Compare this to an asexual female. She will lay eggs that contain exactly the same genetic material as her own. Her offspring will be completely the same as herself. No 64×10^{12} possibilities, just one. That is an enormous difference. Surely, the advantage of sex must be sought here.

Why Sex is a Good Thing

The vast amounts of genetic diversity that are generated by sex have an important effect in the long term: it allows natural selection to operate. Natural selection works by favouring beneficial mutations and weeding out harmful mutations. In an asexual lineage, a beneficial mutation can be stuck in a bad genome without any way of getting out.[9] Its beneficial effect will be cancelled out by the detrimental effects of other mutations with which it is associated and the beneficial mutation will go extinct. By contrast, a beneficial mutation in a sexual population will be in a different genome every generation. It may be in a bad genome in one generation, but in a good genome in the next. Selection then works on the gene, rather than on the genome as in asexuals.

The idea that sexual recombination allows more efficient adaptation was tested by M. R. Goddard et al.[10] in yeast. Yeast is not male and female, but it does have sexual recombination (involving 'mating types'). By deleting two genes that are essential for meiosis and recombination, the authors created a yeast strain that was unable to reproduce sexually. When these asexual yeast cells were grown under harsh conditions (hot and salty) they were slow to adapt. Normal sexual strains increased their fitness over the course of 300 generations. The asexuals also increased in fitness, but at a significantly slower pace.

This ability to adapt to changing circumstances is thought to give sexuals a competitive edge over asexuals in situations where they have to deal with predators, parasites and competitors that are all themselves evolving too. Furthermore, genetically diverse sexual populations should be better at exploiting a given pool of resources. In a heterogeneous environment with many competitors, sexuals will do well by exploiting pockets of underutilized resources – thereby avoiding competition. However, sexuals do not always outcompete asexuals. In harsh environments where mortality is high and resources are not fully exploited, competition is less important. In ephemeral habitats, such as rain puddles, all inhabitants die or move away at irregular intervals, again reducing competition when the habitat reappears. Furthermore, asexuals can outcompete sexuals by utilizing a narrow resource spectrum in homogeneous environments. Asexuals, with their ability of explosive population growth, do well in such environments. In other words, the most favourable reproductive mode for a given environment critically depends on the structure and abundance of resources.[11] This is why we find abundant asexuals in saline lakes, rain gutters and agricultural fields.

Getting Rid of the Garbage

In addition to being inefficient at adapting, asexuals lack an efficient mechanism for getting rid of harmful mutations. A sexual population will generate a variety of genotypes each generation, some heavily loaded with bad mutations, but others relatively clean. The latter will be favoured by selection and produce most of the next generation. The heavily loaded genotypes will die, taking their mutations with them. An asexual lineage on the other hand, will clock up harmful mutations without ever producing less-loaded offspring. Eventually the genetic load will become so large that it kills the clone.[12]

Genetic manipulation of the nematode worm *Caenorhabditis elegans* has shed light on the advantage of sex in terms of shedding mutational load.[13] Populations of *C. elegans* consist of hermaphrodites that can self-fertilize and males (but no females). Normally about 5 per cent of the population has sex in a given generation, while the rest fertilize themselves. The authors constructed strains that were either obligate selfers or obligate outcrossers. The obligate selfers had lost the ability to have sex, while the obligate outcrossers had lost the ability to self-fertilize. As in the yeast experiment, the outcrossers were better able to adapt to stress, this time in the form of a bacterial pathogen. Interestingly, the experimenters also exposed the strains to a toxin that increases mutation rates. The outcrossers had little problem dealing with this. Selfers, however, suffered severely. Normal strains did the smart thing: they increased their rate of outcrossing when exposed to high mutation rates for fifty generations.

Sex and the Genome

The higher evolvability of sexuals and the inability of asexuals to rid themselves of deleterious mutations have been discussed for decades. Even though recent studies are beginning to provide experimental evidence to support them, theoreticians still struggle to explain how these mechanisms could outweigh the twofold cost of sex. Today, many researchers still feel like one of the leading evolutionary theorists of the twentieth century, John Maynard Smith, when he stated, 'One is left with the feeling that some essential feature of the situation is being overlooked.'[1] In this day of genomics and genome sequencing, the time has come to consider whether the 'essential feature' is hiding in the genomes of asexual organisms. We will explore two such possibilities: transposable elements and epigenetics.

Transposable Elements

Transposable elements (transposons or 'jumping genes') are small DNA fragments (1–10 kb), whose structure and copy mechanisms remind us of viruses. They reside in host genomes and have the ability of independent self-replication by virtue of proteins encoded by the elements themselves. Their jumping process is called transposition and often interferes with the proper expression of host genes and causes erroneous chromosomal recombination. This, together with the metabolic burden of replicating and expressing large numbers of elements reduces the fitness of the host. The reason for their persistence at the expense of the fitness of their host is precisely that which distinguishes sexual from asexual reproduction. As we noted above, sexual recombination uncouples a gene from its genomic background. Transposons do well because they copy themselves faster than their host's genome. Their deleterious effects do not affect their own fitness, because the element is in a different genomic background every generation. As such, transposons are 'selfish'. For a detailed review see A. Burt and R. Trivers.[14]

Transposons are found in almost every eukaryote. Our own genome, for example, consists of about 60 per cent transposons. Not all transposons in a genome are active. Of those that are, not all are autonomous. Some transposons have lost the ability to produce functional proteins and rely on hijacking the machinery produced by autonomous elements. Transposons are divided into two major classes. Retrotransposons (class 1) transpose via a copy and paste mechanism. These elements are read by the host replication machinery and translated into a protein that copies the element into another position. DNA transposons (class 2) proliferate by an excision and insertion mechanism. Their proteins chop the element out of the current position and integrate it back somewhere else. For both classes, the target position of transposition is mostly random, which may result in the transposon jumping into a functional gene region.[15]

Sometimes, parts of the host's genome are moved along with the active element by accident, resulting in duplicated genes or control regions. Duplications provide possibilities for novel evolutionary inventions, meaning that transposable elements may also have beneficial effects by speeding up evolution.[16] Furthermore, transposons may be 'domesticated' by the host and for example become new promoters for existing genes. Transposon can thus also have positive effects on the evolution of the host, but this is a by-product rather than the cause. In general, the unchecked transposition of all groups inevitably leads to an increase of elements residing in the host's genome and thus to a build-up of deleterious effects over time.

The Mode of Reproduction and Transposable Elements

Sex, the mixing of the genomes of two individuals, facilitates the spread of transposons. For example, when an individual inherits an active transposon from one parent, but not from the other, the transposon may copy itself to the 'clean' chromosome. On the other hand, sex also provides a means of curbing the spread of transposons. Transposon copies may get lost through recombinational repair if heterozygous, or silenced by control mechanisms (e.g. RNA interference, DNA methylation). Furthermore, selection against highly loaded genomes will also tend to keep the number of copies down (see Figure 3.3).

In asexual lineages, transposon proliferation is restricted to vertical transmission from mother to daughter. Therefore, transposons in asexuals are stuck in the respective host lineage and dependent on that host lineage for its own fitness. In principle, this should select against transposons deleterious to the host. However, as we have seen, asexuals are invariably descended from sexual ancestors and have usually only recently become asexual. The transposons they carry are thus also descendants of transposons that used to reside in a sexual genome and hence have been selected to be active regardless of the fitness consequences for the host. A transposon in a newly asexual genome has no way of 'knowing' that it is no longer in a sexual genome and should thus behave as if it is still in a sexual genome. In other words, it will keep jumping, now unconstrained by some or all of the host's suppression mechanisms. There are two possible outcomes. First, the transposons accumulate in the asexual lineage, eventually creating a heavy genetic load that leads to the extinction of the lineage. Alternatively, the asexual lineage may eventually rid itself of harmful active transposons, if the least loaded clones overcome early extinction. Theoretically, the latter is possible if elements are excised from genomes and effective population size is large.[17]

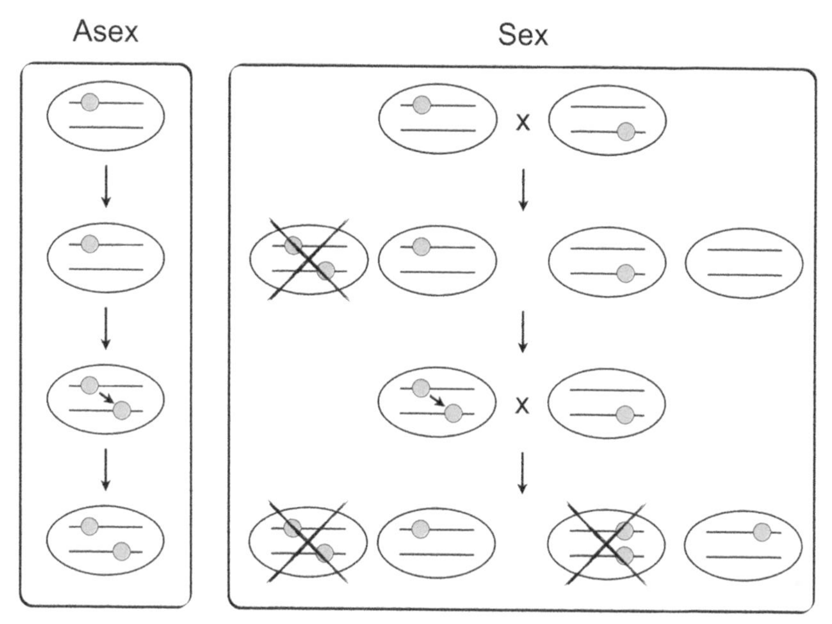

Figure 3.3: How sexual reproduction can help to contain the proliferation of transposable elements. In an asexual lineage, TEs accumulate over time because of unchecked transposition. In sexual populations, TEs will spread and selection will curb the highly loaded offspring. In the first generation, individuals with TEs in one of their two chromosome copies mix their genomes, which results in offspring with two, one or no TE at all. Natural selection will 'delete' the most loaded genome. Transposition in this generation may increase TE copy number, so that both chromosome copies are occupied. Mixing results again in more or less loaded offspring on which natural selection operates. Therefore, sex provides a means of both spreading and curbing TEs from populations.

Currently, the evidence for either scenario is mixed. For example, certain transposons segregate at higher frequencies in the self-fertilizing *Caenorhabditis elegans* and *Arabidopsis thaliana* than in a cross-fertilizing related species.[18] On the other hand, obligate parthenogenetic populations of *Daphnia pulex* appear to have lower copy numbers of both DNA transposons and LTR retrotransposons than cyclic parthenogens of the same species.[19] Along a different line of evidence, a screen for widespread TE types in the putatively ancient asexual bdelloid rotifers failed to find gypsy and LINE elements, perhaps suggesting that bdelloids have managed to avoid early extinction and lost their transposons, allowing them to persist as asexuals for millions of years.[20] However, subsequent work has revealed that bdelloids harbour a considerable diversity of TE types.[21] Clearly, transposon loads should be measured in more asexuals to clarify the situation.

Potential Side Effects of Asexuality

Asexual reproduction has profound effects on the state of the genome, depending on the mode of asexuality. For example, *Wolbachia*-induced parthenogenesis can lead to complete homozygosity, while asexuality achieved through the fusion of two polar bodies results in complete heterozygosity. For the asexually produced offspring to develop as females, further modifications to the sex determination system may be necessary. For example, *Wolbachia* can turn a haploid wasp gamete into a diploid gamete by interfering with the host's cell division, but further manipulations may be needed for the wasp to develop as a female. Sex determination in the parasitoid *Nasonia* involves sex-specific splicing of a gene called *transformer*, which is achieved through an epigenetic mechanism.[22] If *Wolbachia* interferes with this system to induce female development, that may have important side effects. The precise epigenetic mechanism causing sex determination in these wasps is not yet known, but may be methylation. Methylation is a chemical modification of the DNA molecule that influences whether or not a gene is expressed. In some organisms, methylation is also used to suppress transposon activity. If *Wolbachia* removes methylation from the host genome in a non-specific way to induce female development, it might also reactivate silenced transposons.

Alterations of other (components of) sex determination systems may have similar effects. For example, proteins from the Argonaute family play pivotal roles in sex determination in some, perhaps many, organisms.[23] The same, or very similar proteins play a role in the suppression of transposon activity in several organisms.[24] It seems likely that in some cases, the establishment of asexual reproduction involves the disruption of the normal expression of Argonaute proteins. If so, this could interfere with the silencing of transposons, resulting in renewed transposition.

It is thus clear that the mechanism leading to asexuality may have important consequences on the stability of the genome, at least theoretically. Although we understand the cytological mechanism resulting in asexuality in some systems, we know very little about the underlying molecular processes. Identifying these mechanisms and determining their side effects will be a rich area for future study.

Scandalous Asexuals

We have seen that sex is a good thing: it generates huge amounts of genetic diversity, which allows natural selection to operate more effectively. Sex allows an organism to adapt more quickly and prevents the build-up of genetic load. Asexual lineages can be successful in the short term, but tend to become extinct within a relatively short time span. There are a small number of notorious exceptions: the so-called ancient asexuals that are thought to have been reproducing without sex for millions of years: bdelloid rotifers, darwinulid ostracods and several groups of oribatid mites (see Figure 2b). These organisms challenge everything discussed above and were famously termed 'asexual scandals' by O. P. Judson and B. B. Normark.[25] The best-studied ancient asexuals are the bdelloid rotifers, small transparent organisms that inhabit temporary water bodies. Evidence suggests that bdelloids have been living without sex and without meiosis for many millions of years.[26] There is a rich and growing body of literature on bdelloids, a summary of which is outside the scope of this chapter. However, one intriguing piece of evidence that has emerged recently needs mentioning. E. A. Gladyshev et al. found that the genome of bdelloids contains pieces of DNA from completely unrelated organisms, including bacteria, fungi and plants.[27] We do not yet know whether these pieces of foreign DNA are expressed in bdelloids, but the finding opens the intriguing possibility that bdelloids use DNA from the environment as a source of genetic novelty. They could be exchanging DNA between each other in this way too. If so, bdelloids would be having sex after all, just in a very unusual way.

Acknowledgements

We thank Stephan Scheu for valuable comments on earlier drafts of this chapter.

4 EVOLUTION AND ILLUSION

Jagdish Hattiangadi

Illusion, Appearance and Reality

Evolution and illusion are well known concepts. This creates a difficulty in their understanding. When well known, and in no apparent need of clarification, hidden differences are masked in our subtly differing understandings, which may undermine communication. Of the two, it is perhaps illusion that is more in need of clarification on this count.

Illusions are appearances that deceive us. This much is commonly understood. How they deceive us gives two different conceptions of illusion. In modern philosophy, and more generally in modern thought, illusion is taken to be primarily a subjective or a psychological error as to what is out there, a distortion of the senses, or of our perceptual apparatus. René Descartes endorsed a strict dualism of mind and matter. The mechanical universe excludes all non-mathematical, qualitative phenomena, i.e., such things as colours, sounds, smells, tastes, and also such aspects of things as their forms, species and purposes. A mind, within which such qualities are perceived, may find that these 'inner' qualities misrepresent things that are really there in the external world, i.e., in the mechanical universe. Illusions on this account are unreal and subjective. Even if we do not subscribe to Cartesian dualism, we should acknowledge such perceptual illusions. Our interest, however, is in another and older tradition of understanding deceptive appearances and illusions.

Parmenides proposed a theory of motion, or change, in which he concluded that one unchanging thing (that which is) alone is real. That which is not, which includes any change or motion, cannot be or even thought to be. Change is an illusion, a deceptive appearance. Later physicists and metaphysicians, influenced by Parmenides, continued to regard motion as *somewhat* illusory. The unchanging substrate that underlies a changing thing was thought more real than the change. This metaphysical doctrine, widely accepted until the twentieth century, depicts change as manifest in appearance, but less real than the unchanging substrate. This understanding of appearance is of greater interest to us.

In ancient Indian philosophy, illusions that are less than fully real, but nevertheless manifest in appearance, are called 'maya'. We live our ordinary life with our ordinary attachments and values. This existence is described as unreal, in the sense that neither the happiness nor the suffering that we endure, nor the trappings of a normal householder's life, are what our true existence is all about. The underlying reality being hinted at is a religious claim that need not concern us, of course. We notice a different kind of illusion in ancient India in maya than that which is merely a subjective error. Maya is not an illusion in the sense that our everyday life is not at all real, or imaginary; but that it is less real than something else, something hidden, be that what it may.

Arthur Schopenhauer, writing early in the nineteenth century, seems to have been influenced by this notion of a manifest illusion by which we live, and must live.[1] We find this thought also among those whom he influenced, such as Friedrich Nietzsche and Sigmund Freud. Schopenhauer found it interesting that an inner object of our desire, namely, a felt need to relieve a certain special kind of pain or tension, underlies our attainment of pleasure. Pleasure is so organized in us that when we seek it, we are thereby led to do things that satisfy our real need, even if that need is unknown to us. Thus, eating food sustains us, though we can eat merely to satisfy an illusory pleasure, i.e. to relieve hunger or boredom. Similarly, our sexual appetites may be superficially triggered, which can serve to mask the more real need to procreate.

James J. Gibson has noted, taking issue with the purely phenomenal analysis of illusions, how important 'affordances' can also be in understanding illusion.[2] Reality can sometimes afford a look that we mistake for reality. When we set a trap for the live and humane capture of an animal, for instance, we may use arrangements of naturally occurring leaves and branches to disguise the trap into which the animal is lured. Here the illusion of normality that deceives an animal is afforded by the real arrangements of real things in the environment of the trap. Affordances, whenever they are of the kind that deceive us, are what I have called manifest illusions.

The evolution of eukaryotes, which is our concern in this essay, appears to be illusory in this older sense of the word. It is perfectly real at a superficial level, but it is a 'show' or 'maya' that masks the reality that lurks behind it, which is truly the evolution of prokaryotes.

Evolution

Evolution is a gradual change in anything. While that is the general concept, a more particular application is relevant. In current thought, the word 'evolution' is more particularly associated with Charles Darwin's theory of the evolution of species, which is how the word is being used.

When Darwin proposed his theory of the origin of species, he reconciled the apparent existence of species of living forms with the mechanical account of all underlying reality. The universe as a pure mechanism must allow for the existence of kinds, or species. That they exist is manifest to us when we perceive things as belonging to recognizable kinds. But kinds, species, forms are illusory in that their existence is superficial and not fundamental to things, as opposed to what the followers of Aristotle or Plato or the Scholastics had imagined.

The older way of thinking about species is to regard them as more real or more fundamental than motions. In Scholastic metaphysics, to understand the nature of a thing is to understand, among other things, why it moves as it does, or grows as it does. The nature of a thing limits the motions of which it is capable. To understand the cosmos, therefore, is to understand its nature, and the natures of all the various kinds of things within it. Metaphysics, when it is fully developed, would therefore give us a chain of being, where every nature of each being would be found in its place. And the natures of things in turn would explain all change or motion, including in particular all those local motions that may be called 'natural'.

In the new physical theory developed for a moving earth, Galileo Galilei described motions without reference to the natures of the kinds of things moving. The law of falling bodies, for instance, describes the distance a body falls, when free of resistance, in proportion to the square of the time elapsed. There is no reference to the kind of body falling, or to its nature. Descartes based his description of a mechanical universe on laws of motion at a very abstract level, with the understanding that all species, or kinds, or forms, or natures would be derivative entities of undifferentiated matter within this new mechanics. Modern thought has followed Descartes in this, at least until the twentieth century. In the study of life, however, species and forms seem intrusive. This creates a certain tension in the understanding of living things within a thoroughly mechanical point of view. The new understanding of species in Darwin's *On the Origin of Species* reinterprets them to conform with the new mechanical understanding of the universe rather than to the Scholastic understanding.

Darwin provides an account of species as mutable. Earlier, the argument from design was found to be a very persuasive doctrine, whether in Isaac Newton's *General Scholium* or in William Paley's metaphor of a clock. The argument from design points to the existence of an intelligent being who designs the universe, which seems inescapable precisely because species are so well adapted that they could not have been there by mere chance and the laws of mechanics. Darwin describes the adaptation evident in species as the result of natural selection, which acts by culling less adapted, or maladapted, varieties. Small variations arise spontaneously among the offspring of parents in any species; if most of these varieties get culled by the process of natural selection, then whatever remains evolves to produce all future species and higher forms. Those that are culled are likely to be more maladapted than those that are not; and on this simple principle does the apparent adaptation of the species that remain will rest. He envisaged a tree, with many branches lopped off; leaving behind the sharply separated species that we see today, which once began as variations in a single species. Darwin's account is also important in the light it throws on an embarrassing feature of the fossil record: in every era we seem to lose to extinction well over 90 per cent of all the species found in the fossil record of a previous era. This extinction of species on such a grand scale is an embarrassment to any genuine believer's theological account of intelligent creation, but this extinction is to be expected in Darwinian natural selection of mutable species

Explaining the origin of mutable species requires a sufficient supply of small variation in offspring to generate future change of form. Were this variety to disappear, i.e., if offspring became too faithfully similar to the parents and to each other, then evolution would slow down. Explaining why there is such a great variety of forms of life was not easy to understand in Darwin's time. He thought that offspring may combine traits from both parents, and thus produce a blend of their traits. But if this is taken to be the form of inheritance, aptly called 'blended inheritance', there is a serious difficulty with why there is so much variety. Darwin was aware of the difficulty, though he seems not to have known how it may be overcome.

Imagine a large population of plants with flowers of a uniformly white colour. Now let us suppose that a flower develops for the first time a red variation. Let us suppose that bees or other pollinators preferred red flowers to pollinate. Then red flowers would flourish. However, blended inheritance would undermine the success. For the in the first generation, an offspring of a white and a red flowering plant would be pink. In a few generations the increasingly pale pink flowers will be barely distinguishable from white flowers. Without a distinct variation to prefer, the preference for a colour would become ineffective, and population would revert to average.

Gregor Mendel worked out a theory of inheritance that was not based on blending, but on genetic traits that remain discrete units of heredity. His experimental work with peas also led him to the discovery of recessive genes, as opposed to the dominant genes that we may see manifest in offspring. The difference between the manifest phenotype and the genotype that underlies it became an important idea of the new genetic account of populations. Mendelian genetics and mutations were for a while thought to be alternative theories to Darwin's. The combination of the laws of Mendelian genetics and Darwinian natural selection, however, has given us a formidable understanding of how living things have evolved. The 'new synthesis', as it has been called, with its detailed account of population genetics, has given us a robust account of heredity and evolution by natural selection. Moreover, subsequent study of the physical and chemical basis of heredity, in the discovery of the structure of DNA and RNA, in the deciphering of the genetic code, and in the method of DNA sequencing, and much else besides, has greatly enhanced our understanding of evolution.

Mendelian genetics can explain why there is such a variety of living things, and so much variation in displays within species. Discrete units of heredity, i.e., the genes, facilitate the preservation of a variety of manifest form, whereas blended inheritance makes for a dull reversion to the mean. So let us gaze upon the brilliant variety of all living forms and marvel at how little intelligence went into its creation. The new synthesis, with its analysis of populations and 'gene pools', seems to have got it just right.

To conclude this section, let us note then that by the word 'evolution' used here, we should understand the conception of evolution found in the 'new synthesis', with its emphasis on Mendelian inheritance within Darwinian evolution. We will explore two niggling difficulties within this approach. One of these is centrally about the place of sexual reproduction within this Darwinian philosophy that incorporates Mendelian inheritance. A second difficulty concerns the origin of the kinds of organisms within which all sexual forms of reproduction arose.

The Twofold Cost of Sex

John Maynard Smith wrote of the 'twofold cost' of sex.[3] The argument is so simple that it requires little by way of theoretical background. Heritable success in any form of life is manifest in how well the successful form donates its traits to the next and subsequent generations. The more offspring that carry its traits, the more successful it is. If we follow a single trait, then, we can note how successful that trait is by determining its rate of replication in the population from generation to generation. However, this must be understood in a competitive manner: if the rate at which a successful trait is replicated in the next generation is slightly more than that of all rival traits, then the successful trait will overwhelm the population in time. Traits may also exhibit cyclical variations in their replication rate, or they may form a stable equilibrium with rival traits over time.

Sexual reproduction is itself a successful trait in a great many populations. This creates a paradox. If any individual in a population of normally sexual reproducers were to abnormally reproduce asexually, the asexual reproducer would have the advantage of providing twice the number of its own traits to its own offspring within the gene pool when compared with its sexual rivals. If asexual and sexual offspring had equal rates of survival, then those that reproduce sexually would be at a great disadvantage. They would be represented in each subsequent generation at half the rate of the asexual reproducers. Sexual reproduction should soon be rare or extinct in any population that has any asexual reproductive varieties. This is the twofold cost of sex. That sexual forms are often successful and usually obligate is the difficulty that we must face. What is the advantage for survival of a sexual form of reproduction that offsets and overcomes its inbuilt twofold cost? When sex becomes obligate for some species, of course, this question is moot; but another question remains: how does that obligate sexual species survive up to that point of no return, given the twofold cost of sex?

A number of explanations have been offered for the prevalence of sexuality in the face of these overwhelming odds against it. These explanations may be divided into two basic kinds, though there are also hybrid models combining the two kinds. In one basic model, the explanation lies in the organization of the sexual forms, and in another class of basic models the explanation lies in environmental factors that favour sexual reproducers to offset the inbuilt cost.

Those that rely on the compensating advantages gained because of the organization of reproducers, which may be mechanical features or certain statistical features of populations, have the difficulty that these gains are not stable over the generations and across different populations. Mechanisms and populations can arise that are different from those that are conducive to sexual reproduction. The likelihood of unfavourable conditions arising interferes with the explanatory adequacy of the mechanical or statistical model for sexual propagation. Those basic models that rely on environmental benefits to explain the value of sexuality explain too much, because there are many environments in which the advantage is missing. Moreover, asexual species that flourish must do so within the same class of environments. Such accounts of the success of sexual forms of reproduction tend not to be robust for the general case, though for some species and in a limited set of environments, such an account may work very well.

Maynard Smith himself first proposed that there must be some compensating advantage to be gained on one model, as follows.[4] Suppose that there is a 'mutator gene' which creates mutations in other parts of genetic material. It is easy to show that in a bacterium (typically one double strand of DNA), which is celibate (not recombinant), this mutator gene will take over the host: even if a favourable mutation in the genophore is a one-in-thousand chance, after all the rest have died out, the one favourable mutation will give the mutator gene a free ride by fixing it in the population, and the mutator with it. Maynard Smith calls this 'hitch hiking', where the mutator gene gets a free ride derived from the advantage of one of its own advantageous mutation-creations. Sexual dimorphism could be such a mutator that is successful by 'hitch hiking'.

In diploid sexual organisms such an advantage is not stable over many reproductive cycles. Only when it is on the same chromosome, or linked, would a hitch hiker have any advantage by virtue of linkage. Sex therefore can 'hitch hike' a free ride on a favourable gene, by being on the same chromosome and closely linked. When mitosis occurs, chromosomes divide and join in the middle forming crossed pairs of sister chromatids. In meiosis, these structures can also sometimes exchange pieces of chromosomes (cross overs). Cross overs are fairly random and not uncommon. They will eventually separate any two genes, however close they may be on a chromosome, if we wait long enough. In any population it would seem that sex would sooner or later become unhitched with its unknown benefactor long before fixation. It would quickly disappear once it is on its own, because of its twofold cost. Hitch hiking thus cannot explain the stability of sexual reproduction over many generations.

Another resolution to this difficulty is favoured by Michael Ghiselin, following a passage in Darwin's writing that describes the profusion of plants, birds, worms and flying insects in the 'tangled bank'.[5] Ghiselin suggests that sexual reproduction increases variation of offspring. This, in turn, helps a sexual reproducing species to exploit many more subtly different environments than one with fewer variations: it better uses the tangled bank. This advantage is real, no doubt. The difficulty of this offset to the twofold cost of sex is that it is not general enough. It does not explain how it is that there are long periods of stasis (pointed out by Eldridge and Gould) when variation of form is minimal.[6] This would be hard to reconcile with the tangled bank model. There is some evidence of parthenogenesis among some insects during certain periods even when they have the ability to reproduce sexually.[7] Why would they not exploit more environments than less? There are also successful species that support very few offspring in each generation, but which also are sexual reproducers – a strategy that would not benefit as much from the tangled bank, which requires a profusion of varieties to take advantage of available micro niches. These difficulties do not cancel the benefits Ghiselin has stated, but they make them less universally applicable than needed.

Sexual reproduction also gives species 'hybrid vigour'. In asexual reproducers, there is the threat known as 'Muller's ratchet'.[8] In small populations, when deleterious mutations pile up, there may yet be an individual that is missing a very deleterious allele, which would be useful for the population to survive, going forward. Such an individual may be randomly eliminated in a small population, and with it the hope for the future of the population. The mechanism of sexual reproduction, however, encourages the preservation of the good allele in multiple copies in the gene pool of even small populations. So the population has a better chance of survival. We should note, however, that in larger populations this effect of hybrid vigour is much less pronounced. Therefore, this model also does not successfully deal with the general case of sexual reproduction, though in some cases it does provide the benefit it claims.

The one advantage that sexual reproduction confers on any population is that it gives its members more genetic variability from the same gene pool. We have to be very careful here that we are not invoking group selection when we argue that this explains the origin and maintenance of sex. (We cannot say 'and this makes evolution go faster' since this must be shown to lead to fitness in the individual trait that can overtake a population before we can show how it works.) Sexual propagation may be imagined to be advantageous to all those organisms that are in unstable or changeable physical environments. However, a survey of the climatic and geographic distribution of sexual and asexual species by Graham Bell suggests that in fact sexual species are much more common in stable than in unstable or marginal physical environments.

One suggestion along these lines is very interesting. Asexual siblings, being very similar in their heritable traits, are susceptible to mass extinction by parasites. Sexual species will have varieties which can resist a virulent parasite. Sexual species will therefore more likely survive a pandemic.[9] This is a good explanation of an offsetting advantage, so far as it goes, but when there is no pandemic, the allele which is asexual will quickly re-establish itself, because it then has a twofold advantage. Sexual reproduction may have some advantages (during pandemics) but this still fails to explain the general case.

George Williams suggests, moreover, that sexual and asexual forms can fill different existing ecological niches.[10] Citing forms like rotifers and ferns that have both asexual and sexual phases, this seems the better conclusion to draw. However, the origin of sex and its maintenance in every species in the form maintained needs a general answer, which Williams concluded he could not provide; he did not even attempt to make a general case; his studies, however, remain useful in challenging other models of the advantages of sexual reproduction proposed to blunt the evident twofold cost.

Any basic model relying on the environment cannot account for the general case, including the great variety of ways in which sexual reproduction benefits such reproducers, and the different ways in which these traits can be found exhibited in different species. What for instance are we to make of bdelloid rotifers? These are invertebrates living in fresh water that have been reproducing asexually, without any sexual phase, for an estimated forty to eighty million years. Not only have they survived without succumbing to micro parasites, but show a remarkable degree of genetic variation among them, so that they can take advantage of a 'tangled bank'. They seem to be able to survive dehydration for years and high levels of ionized radiation by using a very advanced form of DNA repair. It also seems that they can absorb some DNA from the environment (much like prokaryotes) so that the deleterious traits in their pool do not accumulate.

These criticisms of various models are not recounted in order to deny that the avoidance of parasites, or the exploitation of micro niches, or of hybrid vigour, gives many multi and differentiated celled eukaryotes some genetic advantages because of their sexual reproductive habits. The point is rather that each such model is too particular to be used to account for the general case. In a similar way, we could point to those species that are alternately asexual and sexual in their reproductive habits. Among the Asplanchna rotifers, for instance, there is alternately both asexual reproduction, in the summer, and sexual reproduction, usually when overwintering; but the driver of the choice seems to be weather, and unrelated to the presence of micro parasites. Similarly, Williams cites ferns also among those species that alternate between sexual and asexual reproduction, and though this is also related to environmental factors, the presence or absence of infectious parasites does not seem to be the deciding factor in the choice. For all these reasons, the model of avoiding parasites may be an important advantage enjoyed by sexually reproducing species, but it is not enough to explain the general case, in which so many multicellular eukaryote species are sexually dimorphic.

For large asexual populations, another model has been proposed: there may be multiple potentially deleterious traits, each of which is not lethal, but several together may be. In this case, the value of sexual reproduction is to separate some individuals with many deleterious mutations, and others with few, allowing selection to work faster. This is also a model that works in many cases, but the restrictions placed on the deleterious effects of such mutation, and how often they must happen, severely restricts the cases to which the model applies. Once again, the problem is that the model does not address the general case.

Let us extricate ourselves from these and other models, whose ingenuity is undeniable. Let us acknowledge the detailed knowledge gathered by biologists of the astonishing variety of living forms, which is overwhelming to an outside observer. So where do we stand?

The new synthesis amalgamated the theory of evolution by natural selection with an account of discrete genes that provide the means for producing the astonishing variety of flora and fauna that we witness, on which natural selection can usefully operate. Without the new synthetic modification of it, Darwinian theory would have been defeated by a statistical reversion to the mean.

However, the new synthesis comes with its own cost: Mendelian genes work by a logic of their own. It is not enough to show that a certain trait has benefits, it must be followed up by which individual traits it benefits, and whether these very individuals within the gene pool will differentially survive as a result. The general case for sexual reproduction is that it benefits the species; but the twofold cost should ensure that every sexual trait would be replaced from the population by asexual varieties, in time. Ingenious mechanisms for sexual benefits are met by equally ingenious analyses of situations in which the mechanisms may fail. Ingenious effects of the environment are proposed, but the very variety of living forms and the ways in which they exploit sexual reproduction shows that any particular model of environmental factors is never general enough.

If we seek a general answer to this question it must be one that shows how all these models can be used together, to explain how living forms can exploit the advantages of sexual reproduction without losing out to its twofold cost. This leaves us with an intriguing question. Which genes do not pay any of the twofold cost of sexual reproduction?

The Origin of Eukaryotes

When we face a difficulty in a received and valuable theory in a science, the proper use of inductive method is to avoid trying to solve the problem, or resolve the difficulty, all by itself. For if we try, we find that there are far too many options available to know what to do. The proper use of the method of induction is rather to generate more difficulties of a different kind for the same theory; and then to consider all the difficulties together, as if it is one riddle or puzzle.[11] The puzzle consists, in such cases, of multiple intractable problems as its nodes. Then the solution to the problems simultaneously forming the puzzle will fit the underlying facts very tightly, just as a solved jigsaw puzzle fits all its entire component pieces together. The advantage of this inductive strategy is that the variety of answers possible to one problem will be mostly blocked by the peculiarities of another problem. If there are enough problems forming the nodes of a puzzle, then the puzzle will have a unique solution that yields the model that we seek.

Sexual reproduction governed by Mendelian laws of inheritance is found in multi-celled eukaryotes with differentiated cells. The new synthetic theory cannot be applied in any straightforward manner to the origin of the eukaryote cell itself, which began once as a single cell. However, we could argue that whatever we have learnt about genes may still be applied to prokaryotes and early eukaryotes (single celled protists) even if they were not yet sexual in their reproductive habits. Here we must pause to digress a little.

The genotype theory of heredity is now quite well confirmed as an experimental fact within biology. However, it has been suspect in social and political philosophy for the strong individualism that it implies.[12] It is suspect, perhaps, because of its implications for human society, when it is somehow generalized to apply to it. This has led suspicious researchers to seek an alternative to the new synthetic theory even within biology, seeking, for instance, more place for cooperation as a basic principle among living things, with a correspondingly reduced role for competition. Among the strong contenders for a cooperative theory of evolution is an account of evolution partly based in symbiosis. No biologist would want to deny the evidence of extensive symbioses in the history of biology. However, could we make cooperation a basic principle of evolution, as we have made competition among heritable variations?

Early in 1970, Lynn Margulis published a pioneering book reviving the origin of organelles found among eukaryote cells in the endosymbiosis of some prokaryote organisms.[13] Margulis brought a great deal of diverse factual evidence to bear on it. The sheer quality of factual evidence that she was able to marshal made it hard to dismiss her initially unpopular suggestions out of hand. Although some of her conjectures concerning the origin of some particular organelles have not been widely accepted, her main thesis, namely, that chloroplasts and mitochondria, which are organelles in the eukaryote cell, seem to be derived from free living prokaryotes, is now well supported by more recent sequencing evidence. In particular, the mitochondrion within the eukaryote cell is widely acknowledged to be a residual α-proteobacterium, while the origin of the nucleus is widely acknowledged to be derived from a member of archaea (or archaeabacteria.) Chloroplasts, which are found in plants, but not in all eukaryotes, seem to have had a facultative cyanobacterium at its origin.

Margulis herself, since publishing her first work, suggested that her ideas are 'a change in thought style' as Ludwick Fleck expressed it, or as a 'paradigm shift', using a combination of words each of which was made popular in philosophy of science by Thomas Kuhn. Here is how she sees the difference in the style of thought:

> It is that ultimately males and females are different from each other not because sexual species are better equipped to handle the contingencies of a dynamically changing environment but because a series of historical accidents that took place in and permitted the survival of ancestral protists.[14]

Margulis's suggestion is that the eukaryote cell is a product of a series of cooperative accidents in nature – of 'serial endosymbiosis'.

Any change in 'style of thought' leads us to ask what benefit it brings to a better understanding of the subject. In the standard new synthetic account of the origin of any traits whatsoever, there are always certain irreducible accidents that are admitted. Mutations and crossovers are random events, for instance, and they are admitted as such by theorists before they go on to study fitness. Even if we happen to be determinists and conclude that all chance is merely apparent, we cannot deny the appearance of chance in the variability of offspring. To that extent at least we must acknowledge it. However, only a small percentage of forms of life survive, as the fossil record indicates. Hence, any appeal to the existence of accidents and happenings, including those proposed by Margulis, must be followed by an analysis of why the new and accidentally formed traits were not so deleterious that they have survived over time. The accidents may account for the variation that there is to be selected for or against, but the survival must depend upon the differential adaptive benefit (or at least a lack of net deleterious effects) exhibited by the variants that survive.

Such an evolutionary explanation may take one of two forms: we could suppose that a complete account can be given for the selection of all the traits that exist in each species on the basis of their adaptation (an approach that is also called 'selectionism' by its opponents); or alternatively, we may allow for a good deal of 'genetic drift' and 'neutral evolution' so that the existence of some varieties and species can be attributed to factors that include an element of chance (a 'neutral theory' of the evolution of some traits). In smaller populations, of course, the element of chance will play a proportionately larger role in determining outcomes. Chance nevertheless appears to play some role in any account of evolution. But chance, though a needed ingredient, cannot by itself account for adaptation. Adaptation must be still explained in terms of natural selection. No appeal to a history of accidents can substitute for an explanation of adaptation by natural selection. The recognition of accidents is much more useful in any Darwinian model for the generation of variations among offspring. This, therefore, is how a history of symbiotic 'accidents' should be interpreted.

Margulis's 'endosymbiotic' hypothesis fits well within a classical Darwinian scheme if there were a number of historical accidents that failed, but permitted the survival of some lucky protists because of the adaptive benefits they obtained as compared to the rest. Then we can examine why sexual species, which emerged among their descendants, are better equipped to handle the contingencies of a dynamically changing environment.

In the basic arrangement of a eukaryote cell, there are several membrane clad organelles. The most important is of course the nucleus, which houses the chromosomal DNA. The two most prominent among the rest of the organelles, the mitochondria and the chloroplasts, also possess their own DNA. The chloroplasts are only found in some of the eukaryote cells, and therefore concern us much less in this essay. All eukaryotes possess, with a few exceptions, the nucleus and the mitochondrion, both of which house DNA.

The presence of a membrane clad nucleus with its chromosomal DNA in the eukaryote cell is a matter of definition: we would not call a cell a 'true nucleated' cell (or a 'eukaryote') unless it had a 'real' nucleus, namely one that is encased in its own outer membrane. The nucleus is the 'star' organelle of the eukaryote cell. It houses almost all the chromosomal DNA within its walls. When there is mitosis, or the cell division characteristic of this kind of cell, it is the chromosome in the nucleus that must divide in two for the two daughter cells to form. When there is meiosis, it is two haploid cells (the gametes) whose nuclei typically fuse to form a unit for the benefit of a new diploid eukaryote. All the proteins are also manufactured from structures within this organelle, and hence all cytoplasm. In a certain sense, almost all the inscribed instruction for a living cell derives from its nuclear DNA.

Mitochondria are found in almost every type of eukaryote cell, with very few exceptions. They are generally referred to as the 'power generator' of the cell, because it is the location in which oxidative phosphorylation (respiration) is centred. In the exceptionally few eukaryotes that lack mitochondria, there is evidence of an identifiable mitochondrial type of DNA that has strayed into the nucleus, which suggests that the amitochondriate eukaryote once had and then lost its mitochondria ancestrally. Since every known eukaryote that lacks mitochondria appears to have possessed this organelle ancestrally, we may conclude that the very first eukaryote cell had both mitochondrial and nuclear DNA from the very beginning.

The mitochondrion is the main 'power generator' within the cell that uses oxygen and nutrients to convert adenosyne diphosphate (ADP) into adenosine triphosphate (ATP). ATP is used in all cells to store energy. When ATP is broken down into ADP and Pyruvate, it releases a controlled amount of energy for use when needed in the cell for its normal functioning.[15] Among other functions of the mitochondria are also the controls of programmed cell death (apoptosis) including the regulation of the numeric limit of cell divisions, and the supply of materials for the manufacture of nucleic acids for the cell.

Two difficulties for this theory of the origin of eukaryote cells remain. One concerns how the 'endosymbiosis' could have happened in the first place, and the other how the joint cell became a stable form of life.

Margulis's first suggestion of a symbiosis of what we would now identify as archaea and α-proteobacteria was thought to be an act of phagocytosis. A primitive amitochondriate eukaryote was thought to have eaten a bacterium that remained lodged thereafter in an undigested form. That sexual differentiation is the eventual product of serial indigestion is a provocative hypothesis, but one with significant difficulties. Phagocytosis seems to be absent entirely among prokaryotes, and appears to have developed independently many times only somewhat later in eukaryote evolution.[16] Yaacov Davidov and Edouard Jurkevitch, however, have given a different but very plausible account of the origin of the multiple DNA in the cell: the original eukaryote was formed by an infection of archaea by α-proteobacteria.[17] There are apparently numerous known instances of prokaryotes infecting and residing inside other prokaryotes. Some α-proteobacteria are also among such infectious bacteria. The evidence also suggests, it seems, that there are bacteria that make infection of other prokaryotes their means of survival. Any α-proteobacterium of this kind would likely have invaded and infected many different types of susceptible prokaryote cells. Out of all those infections, only one formed a stable symbiotic relationship that resulted in the form we can now recognize as the ancestor of all eukaryotes. This interpretation fits a Darwinian model of adaption by natural selection.

For any heritable symbiosis to work, there must be mutual benefits and also no lethal consequences to the joint cell. The main benefit that the ancestral α-proteobacterium brought to the union is the prolific production of chemical energy in the form of ATP, at a rate of about eighteen times the rate of production in the glycolitic metabolism of the archaeon it infected. If we assume that the symbiosis was based on an exchange of sulphides, the archaeon may plausibly be supposed to have contributed its waste product, in the form of sulphides, as an abundant raw material for the benefit of the invading α-proteobacterium.[18] There are other hypotheses that are also interesting, such as the hydrogen hypothesis: the origin of peroxisomes is of relevance, though, because it may show how an anaerobic ancestor of the archaeon host could have become tolerant of oxygen and hydrogen.[19]

In order for this aggressive form of infection to survive – that is to say, for the new combined cell to reproduce as one cell – there must be at the very least a timing mechanism to co-ordinate cell division. Since the protomitochondrion produces a copious, even an excessive, amount of usable energy for the benefit of the united cell, the infected arhaeon will be well nourished. Whenever prokaryotes have plenty of usable energy, they multiply rapidly. This destabilizes the reproductive relationship with the protomitochondrion. The invading α-proteobacterium must therefore have arrived with a well developed predatory strategy of genetic control to slow down cell division within its archaeal prey. This hypothesis should be testable: infectious forms of α-proteobacteria must possess some forms of genetic interference of their infected prey even today. In most infected cells, however, this likely provides only a temporary control, lasting long enough to allow the predator to live off the prey for a while, before it leaves for another host. In just this one particular archaeon, it would seem, this bacterium's method of controlling the prey DNA was successful in becoming a stable structure. It achieved this effect, it seems, by enclosing the host's DNA within a cell wall of its manufacture (this pathway itself being, perhaps, a variation of its own form of cell division in which a cell wall forms down its own middle). The particular archaeon that was the host cell would have accidentally responded in this unexpected way to contribute to a stable reproductive cell. It presumably became a reproductive captive. Its own ability to produce cell walls was disabled and compromised, giving rise to the typical flexible wall of the eukaryote cell.

We should notice here that nuclear DNA appears to be controlled by its surroundings, whereas the mitochondrial DNA appears to be free. Witness the fact that mitochondria reproduce by the simple method of binary fission, as wild bacteria do in their natural state. Nuclear fission in the eukaryote cell, by contrast, is a complex and controlled process of mitosis.

When we posit that an α-proteobacterium first invaded an archaeon and then, over time, sent most of its DNA and many of its functions over to the nucleated cell, the main difficulty that we need to work out is how this transfer could have happened without loss of function. William Martin raises this question poignantly.[20] DNA transfer is accomplished bit by bit, but the transfer is functional only when it is a transfer of entire pathways. All protein production in the free living α-proteobacteria, for instance, has since been transferred from the mitochondrion to the nucleus. How could this happen without temporary but lethal forms of dysfunction? Moreover, it has happened many times over for different such pathways. The nuclear DNA of any eukaryote has by now become a combination of archaeal DNA for regulating information (replication, reproduction, transcription, and so on) and α-proteobacterial DNA for production of the lifestyle of the united cell.

William Martin suggests a reasonable solution. Ongoing errors in protein targeting, combined with the existence of multiple copies of the mitochondrial DNA, would allow copying to take place slowly and gradually, without deleterious consequences, until the pathway was later activated after all the proteins have been found transferred. In an alternative model, the DNA of a pathway can be copied bit by bit into the 'junk DNA' in the nucleus, but activation alone will be a unique, later event. This model can be tested: almost entire sequences of current pathways must also be already copied, but inactivated, in the junk DNA of the nucleus. There must be redundancy in the copied functional DNA of the mitochondrion, of course, in order to allow for transfer without loss of function. We know that there are typically many mitochondria in each eukaryote cell. This requirement of redundancy also explains why the reverse process does not readily occur, i.e., DNA transfer from the unique nucleus to the many organelles.

Eukaryotes as Manifest Illusion

It was not so long ago that the 'two kingdoms of life' were understood as animals and plants, a division of life that may be traced to ancient times. Bacteria were not at first noticed, and when they were, they were identified in the form of infections among animals and plants. Even today, what we know most about bacteria is how they infect us, causing disease and discomfort. It seems ironic in retrospect that we, in the kingdoms of plants and animals, are actually a product of mutual microbial infection.

Whether we take the problem of the twofold cost of sex, or the problem of the origin of functioning multikaryote cells, the inevitable conclusion must be this: eukaryote existence is basically a mitochondrial form of life. Mitochondria are a particular way in which certain purple sulphur α-proteobacteria found a way to propagate themselves. They found an archaeal host with the right responses. This thought can shed light on the twofold cost of sex.

The twofold cost of sex can be shown to be offset by many benefits, not all of which were even listed. All of them, considered together, confer useful benefits on sexual reproducers – the most obvious of which are hybrid vigour, resistance to infection and an augmented ability to exploit the environment. We should note, however, that the twofold cost of sex is never a cost to the mitochondrial DNA. Its twofold cost is a tax levied only upon the chromosomal DNA. Mitochondria benefit from the competition between sexual reproducers, when it contributes to hybrid vigour, without incurring any loss of its own. The ability to fend off infection from other bacteria during threatened pandemics is just another occasional advantage of sexual reproduction over the unprotected life of bacteria in the cruel outside world.

Mitochondria provide copious energy to the symbiotic union. They donate the energetic bounty of the Krebs cycle to a cell that otherwise used glycolytic fermentation. This added energy from respiration, however, is denied to the unified cell for rapid reproduction, which would have been its natural effect. The added energy is utilized instead to allow the united cell to grow large and complex in a manner that no prokaryote can manage on its own. The typical eukaryote cell is as large as a small town or a large village of microbes when compared to a typical prokaryote cell. What the mitochondrion obtains in return for this bountiful energy it provides to the large cell is a protected lifestyle. Ensconced in the cytoplasm of the eukaryote, it is no longer threatened by laterally transferred DNA from many other prokaryotes, which is otherwise very common. It is no longer buffeted by variable chemical environments, most of which would be threatening. It has instead the luxury of a home with a predictably controlled environment that caters to its unique and dependent lifestyle. The means of living its lifestyle, which was once encoded in most of its functional DNA, is gradually transferred to the captive nuclear DNA of the union. Only the minimal reproductive and informational apparatus of the nuclear chromosomal DNA is retained – in effect, allowing the union to produce, by a controlled process of mitosis, more and better housing over time for its own infectious form of life. Unlike most housing, which depreciates over time, this housing is renewable.

When we look around on earth we see such beautiful plants, animals, algae and fungi that it can take away our breath. It is exhilarating to see the diversity and marvel at all the adaptations. Struck by all this, we come up with a theory of evolution by natural selection; well, Darwin does, at any rate. We may be tempted with him to think of all this forming a tree of life.

Yet we are investigating this biosphere backwards – not in its own temporal order. Our investigation started late, and we proceeded by studying what was of greatest importance to us, latecomers as we are. Evolution by natural selection is truly that of the microbes, though their forms of inheritance are messy compared to the form invented by the mitochondrial infection. They do not form a tree. Laws of inheritance are less well structured among the microbes than we have learnt from Gregor Mendel. We have yet to see their pattern fully and to understand how they have evolved. The well ordered heritable forms of life that we studied first in the new synthetic account are illusory. They are a kind of manifest illusion. The history of life is really the history of microbial life.

5 EVOLUTIONARY THEORY, CONSTRUCTIVISM AND MALE HOMOSEXUALITY

Pieter R. Adriaens, Andreas De Block and Lesley Newson

Introduction

Evolutionary explanations of human male homosexuality are sometimes charged with being essentialist, in that they assume such sexual orientation to be a deep characteristic of the individual – a characteristic variously described as 'innate', 'immutable' or 'discrete'. Social constructionists, by contrast, argue that grouping individuals by means of their sexual orientation is rather arbitrary. Traditionally, evolutionary scientists and social constructionists have happily ignored each other at best, and ridiculed each other at worst. Biologists have laughed down constructionism as hypocritical,[1] while even mild critics characterize it as a 'position that lots of people have a hard time understanding'.[2] Evolutionary views of homosexuality, in their turn, have been dismissed as half-baked, anachronistic and 'naïvely insensitive to political connotations'.[3]

However, both evolutionary and constructionist positions come in many guises, some of which are considered to be less extreme than others. In this chapter, we will flesh out some basic claims of the constructionist position and present three evolutionary approaches to homosexuality that go some way to incorporate these claims. The first approach builds on the notion of phenotypic plasticity in contemporary evolutionary biology, and suggests that same-sex sexuality could be a conditional adaptive strategy that enables individuals to strengthen alliances with non-kin. As such, this approach dovetails with one of the pet topics in the social constructionist literature: the historical and cross-cultural diversity of same-sex sexual behaviour. The second approach stems from ideas that have been developed in recent theories of cultural evolution and gene-culture co-evolution. It argues that such theories can help us understand how biologically evolved behaviours cause human populations to share cultural information and pass it down the generations so that cultural information evolves over time in a way somewhat analogous to the evolution of genetic information.

This can result in the emergence of behaviours, including sexual preferences, rather different from those that would emerge from genetic evolution alone. The third and final approach offers an evolutionary explanation for the tendency of both scientists and lay people, including many homosexuals, to take an essentialist attitude towards human homosexuality.

Taken together, these three approaches instantiate and substantiate recent rumours about the project to develop an *evolutionary social constructionism*.[4] By taking seriously claims from both evolutionary theory and social constructionism, we want to suggest new approaches to evolutionary research on human homosexuality. To date, this research has been obsessed with what we deem to be a non-problem, i.e., how 'gay genes' have managed to escape natural selection, which is based on the erroneous assumption that the same-sex sexual behaviour observed in contemporary Western European and North American populations has occurred throughout human evolutionary history.

In fact, during the last two hundred years or so, the reproductive behaviour of most human populations has changed dramatically. This includes sexual behaviour, and in many populations, especially in the West, changes in sexual behaviour between members of the same sex have been particularly conspicuous. Taking into account these changes, we use the expression 'modern homosexuality' to refer to the kind of same-sex sexuality that emerged and came to dominate the Western world in the past couple of centuries. When speaking about 'modern homosexuals', we refer to the mostly Western men who tend to have sex with men (and only men) of roughly the same age and social status, and, most importantly, who identify themselves as 'homosexuals'. 'Same-sex sexuality', by contrast, is taken to refer to a much wider variety of sexual activities between men, both past and present. The real challenge, then, is to explain how and why modern homosexual behaviours and beliefs about these behaviours, which appear to be novel, emerged at this point in human evolutionary history. In this chapter, we want to explore how evolutionary theory can help us understand the origins and implications of this evolution.

What is Constructed about Human Homosexuality?

Ian Hacking once said that the words 'social construction' 'can work like cancerous cells. Once seeded, they replicate out of hand.'[5] Still, some basic meanings if the phrase can be inferred from the abundant constructionist literature. Often prompted by a desire to raise consciousness about something, social constructionists hold that that something is contingent on local social forces. Based on this minimal, and rather trivial interpretation, we are all social constructionists about many things, ranging from language to scientific theories. A much stronger interpretation would be that something is socially constructed not only if it is affected or shaped, but also if it is *brought into existence* by our theories about that something – whether they are folk beliefs or scientific hypotheses. There are not many uncontroversial cases of this kind of social construction, but the history of psychiatry proves to be an interesting source. In *Pharmaceutical Reason* (2005), for example, A. Lakoff described the great lengths to which international biomedical companies went to introduce American mental disorder categories in Argentina.[6] Their products were not selling in Argentina because the diagnoses that indicated the use of these products were simply not being made by local psychiatrists. Argentine mental health care has always been a stronghold of psychoanalysis, and psychoanalysts are known to assess and treat illnesses in ways that are very different from more biomedically oriented psychiatrists. To open up the Argentine market, the biomedical companies sponsored training events, gave free drug samples and distributed educational materials. In doing so, they succeeded in transforming not just how Argentine mental health professionals think about mental illness, but also how their patients experience the illnesses they suffer. Saying that bipolar disorder and major depressive disorder were socially constructed in Argentina means that these illnesses were brought into existence by the deliberate introduction of a new way of practising psychiatry.

Unlike Lakoff, many social constructionists are often unclear about who is doing the constructing, what this constructive 'shaping' or 'creating' amounts to and, more importantly, what exactly is being socially constructed. Regarding modern homosexuality, there are at least three possible objects of social construction: the behaviour of modern homosexuals, their sexual preference, or their sexual orientation. The difference between these three levels is a difference in volatility, with 'orientation' referring to a stable and enduring internal preference for same-sex sexual activities. When claiming that modern homosexuality is a social construction, which of these levels are constructionists referring to? The issue is obviously important, since scholar X can be a constructionist about sexual orientation, but not about sexual behaviour, whereas scholar Y might defend exactly the opposite view.

Minimal constructionists claim that, both historically and cross-culturally, different social conditions have left their mark on the sexual behaviour of individuals. The socio-cultural differences between, say, Renaissance Italy and contemporary Britain – differences as diverse as conceptions of manhood, social geography or scientific beliefs – help us explain the different expressions of same-sex sexuality that we find in both contexts. This kind of constructionism is in fact compatible with some kinds of essentialism, as it allows for the possibility that modern homosexuality is not essentially different from other kinds of same-sex sexuality. Strong social constructionists go one step further. They admit that men have *behaved* in ways regarded as sexual with other men from time immemorial. During limited periods of their lifetime, some of these men may even have *preferred* to have sex with men. Yet the very idea of a homosexual *orientation*, they say, is radically new, in that it involves a new kind of individual, who feels, thinks and behaves in ways that did not exist before.[7] The modern homosexual was a new kind of individual because he was brought into existence by a cocktail of psychological and biomedical convictions, rapidly expanding urban areas and various other social forces unique to eighteenth-century and nineteenth-century Europe.

Michel Foucault, one of the intellectual fathers of social constructionism, was the first to highlight the importance of this transformation from same-sex sexuality to modern homosexuality. In *The History of Sexuality*, he writes that 'homosexuality appeared as one of the forms of sexuality when it was transposed from the practice of sodomy into a kind of interior androgyny, a hermaphroditism of the soul. The sodomite had been a temporary aberration; the homosexual was now a species.'[8] Following Foucault, contemporary sex historians have set out to describe how, as a new species or kind, modern homosexuals differ from their predecessors.[9] First of all, modern homosexuals emerge as a new species because, for the first time in the history of same-sex sexuality, they organize themselves into groups that share members-only meeting places, cruising areas, codes of conduct and various mannerisms.[10] Secondly, members of this new kind date exclusively with men of about the same age and status. They rarely have sex with women, adolescents or seniors. In various past eras and non-Western societies, by contrast, a male might have moved from being loved as an adolescent by an older man, to loving young men himself, and then finally to marrying a woman. Further, many modern homosexuals are overtly effeminate, while in earlier cultures males engaged in same-sex sexuality to enhance their manliness. G. H. Herdt, for example, has documented how the Etoro from New Guinea require adolescents to ingest the semen of a senior group member in order to achieve manhood. Initiating young men into all sorts of manly virtues seems indeed to be a recurring theme in the history of same-sex sexuality.[11] Finally, for modern homosexuals, sexuality is more than an informal action pattern. It becomes a keystone in defining their identity – a process variously described as psychologization, subjectification or interiorization of sexuality.[12] Taken together, these and other differences seem to legitimize the view of strong social constructionists that modern homosexuality is a radically new kind of same-sex sexuality.

Contemporary historians of sexuality disagree with Foucault, however, in explaining how this evolution came about. Many of them fault Foucault for overestimating the influence of biomedical theories in the construction of modern homosexuality. In their view, nineteenth-century psychiatry and sexology simply served to validate a pattern of same-sex sexuality that had already been developing since the beginning of the eighteenth century. By conceptualizing homosexuality as a disease, and speculating about its physical correlates and causes, medical science strengthened the view that modern homosexuals did indeed make up a new species – a natural category to be distinguished from its (equally new) counterpart, heterosexuality. If, however, the biomedical sciences did not invent homosexuality, who did? Social constructionists have come up with various answers to this question. Some have emphasized the importance of societal expectations in shaping a new social role – the role of the homosexual – which is hypothesized to help safeguard the masculine identity of the male heterosexual majority.[13] Others have focused on 'the availability of social insurance other than family support and of sufficient housing stock, at least some of which families do not control', thus suggesting that changes in family life may have contributed to the emergence of modern homosexuality.[14] In the following two sections, we will argue that some theories in today's evolutionary sciences allow us to reconceptualize, i.e., to naturalize and complement, these constructionist claims.

Plasticity and the Alliance Formation Hypothesis

In his exploration of evolutionary social constructionism, David Sloan Wilson emphasizes the importance of phenotypic plasticity, i.e., the ability of a given genotype to produce different phenotypes in different environments.[15] In Wilson's view, examples of such plasticity are legion in the animal world:

> [C]aterpillars that look like twigs in spring and leaves in summer, fish that grow streamlined bodies in the absence of predators but flattened bodies in their presence to exceed the gape of their jaws, frog eggs designed to hatch prematurely at the approach of a snake, salamanders that morph into big-jawed cannibals when food becomes short, and on and on.[16]

Phenotypic plasticity explains why humans, too, have the ability to behave differently in different environments. J. Belsky, for example, has argued that different family environments can have differential effects on age at menarche and pubertal development in girls.[17] Similarly, the human fetus is known to adjust its morphology and physiology to the nutritional state of the mother.[18] These examples teach us that genes should not be considered as blueprints, but rather as intricate sets of if-then rules which enable the organism to key its physiology, morphology and behaviour to ever-changing environments.[19]

A key concept in human behavioural ecology, phenotypic plasticity underscores the social constructionist adage that the potential for change is a crucial part of human nature. The question remains, however, how this biological concept can be made relevant to specific social constructionist claims about modern homosexuality. A recent hypothesis in evolutionary theory, the so-called *alliance formation hypothesis*, may be of help here.[20] Basically, the alliance formation hypothesis of same-sex sexuality holds that same-sex sexual behaviour was selected for in many animal species, including humans, because it establishes, maintains and strengthens strategic alliances between non-kin. These alliances enhance the individual's chances of survival and reproduction, relative to rivals who are less inclined to engage in such sexual behaviour. The alliance formation hypothesis about same-sex sexuality has been shown to hold true for a variety of non-human animals, including bottlenose dolphins, greylag geese, Sumatran orangutans and baboons.[21] Studies on social primates have shown that low-status males often negotiate their position in the group hierarchy by means of same-sex sexual activities, ceasing such activities as soon as they achieve higher status. In langur monkeys, for example, individuals belonging to so-called bachelor bands engage in same-sex sexual behaviour before conducting a collective raid on the dominant male's harem. According to one study, 95 per cent of all sexual activities in a langur bachelor band are same-sex activities, while dominant males have sex only with females.[22]

Human males may likewise use same-sex sexual behaviour to form bonds and alliances in situations where cooperation is vital. Many examples from the historical and ethnographical literature on human same-sex sexuality seem to confirm the basic predictions of the alliance formation hypothesis.[23] Firstly, same-sex sexual behaviour has often facilitated bonding between males. Even though humans differ from other animals in that they have powerful nonsexual bonding mechanisms, such as shared language and rituals, sexuality remains one of the most powerful symbols of loyalty and affiliation.[24] In populations as diverse as the Mamluks of medieval Egypt, the Samurai in pre-industrial Japan, and today's Sambia of Papua New Guinea, sexual behaviour often served to solidify long-lasting companionships that proved particularly important in dire circumstances, such as wars or long expeditions.[25] Secondly, same-sex alliances demonstrably support the survival and reproduction of both partners. Rocke's detailed discussion of same-sex relationships between men and young adolescents in fifteenth-century Florence provides a good example of this mutuality. Although such relationships were socially asymmetrical, they often had important benefits for both parties. Parents encouraged and dressed up their sons to attract older males, so the latter would shower them with gifts and money; and while the adolescents were in it for the money, men needed them to function as status symbols. Indeed, the chances of getting hold of important civic offices crucially depended on the quality of one's 'boy(s)'.[26]

The alliance formation hypothesis highlights just one of the many evolutionary functions of same-sex sexuality.[27] Therefore, it can only account for a small portion of same-sex sexual behaviour in any given species, including humans. Yet if same-sex sexuality is indeed important in cementing strategic alliances between human males, the alliance formation hypothesis provides an evolutionary explanation that is rather congruous with certain elements of the social constructionist position. For example, adherents of this hypothesis consider same-sex sexuality as a conditional strategy, rather than an immutable trait, thus accounting for the constructionist claim that human cultures have varied, and still vary, widely when it comes to same-sex sexuality. In considering the variability of same-sex sexuality as an example of phenotypic plasticity, the alliance formation hypothesis can also account for the increasing occurrence of *exclusive* same-sex sexual behaviour – one of the main characteristics of modern homosexuality – in a rapidly urbanizing Europe. European cities have included subcultures in which same-sex sexuality was common as early as the sixteenth century, and that need not surprise us.[28] For those who spend most of their time in densely populated areas, rather than traditional, kin-based communities, creating and confirming new alliances with 'strangers' may be of special importance.[29] The alliance formation hypothesis predicts that urban anonymity and concern with non-familial pursuits, much like wartime conditions, will foster same-sex companionships.

In considering same-sex sexuality as a manifestation of phenotypic plasticity, the alliance formation hypothesis is very much at odds with sociobiological views on the topic. In our view, some sociobiologists and evolutionary psychologists erroneously assume the men who call themselves 'homosexuals' in contemporary North America and Western Europe to be representative of the entire evolutionary history of same-sex sexuality.[30] This view is apparent in their understanding of same-sex sexuality as an evolutionary paradox – if 'homosexuals' do not have offspring, how are 'gay genes' being preserved by natural selection?[31] Even disregarding the issue of whether such things as genes predisposing men to same-sex sexuality exist, and how we should conceptualize the way they relate to people's sexual orientation,[32] the problem with this view is that, firstly, twenty-first century Westerners are a very small sample of humanity. Therefore, they are unlikely to exhibit more than a tiny fraction of the total human behavioural repertoire. Secondly, when one compares results of psychology experiments obtained with participants from a range of cultures, those obtained with Western university undergraduates are often found to be outliers.[33] The same holds true for modern homosexuality. Historical records and anthropological research show us that modern homosexuality is but one of myriad forms of same-sex sexuality, and it is certainly not the most common one.[34]

The alliance formation hypothesis dovetails with a number of social constructionist claims about same-sex sexuality and modern homosexuality, and can therefore be seen as a genuinely evolutionary social constructionist hypothesis. Social constructionists would probably fault it, however, for concentrating too much on sexual *behaviour*, while ignoring the construction of cultural *norms* regarding such behaviour. Phenotypic plasticity may well be an important factor in explaining the remarkable flexibility of the human species, but there are many other factors to reckon with. In fact humans are more flexible than any other animal species because they rely to a large extent upon innate complex social learning mechanisms – mechanisms which, like genes, have been preserved by natural selection.[35] The alliance formation hypothesis predicts that any environment necessitating the use of alliances will increase the occurrence of same-sex sexual behaviour, regardless of what people believe about such behaviour. However, a society's beliefs about same-sex sexuality do play an important role in understanding same-sex sexual behaviour in that society. Kirkpatrick admits as much: 'humans are quite plastic in conforming to social institutions. In some societies of Melanesia, in 17th-century Japan, and in classical Athens, men have been expected to find men sexually attractive, and on the whole they have done so.'[36] Social constructionism emphasizes how knowledge and concepts constitute people, and it is unclear how the alliance formation hypothesis could contribute to our understanding of how this happens. There are other evolutionary approaches, however, that shed a light on this issue. One such approach is discussed in the following section.

Deconstructing the Power of Families and Elders

An evolutionary approach known as *dual inheritance theory*[37] emphasizes the effect of social learning and social influence on human behaviour and may therefore provide another basis for evolutionary social constructionism. Evolutionary scholars taking this approach have suggested the *kin influence hypothesis*[38] as an explanation for the rapid change in reproductive norms, including the emergence of modern homosexuality, which accompanies the process of economic development. According to dual inheritance theory (DIT), the evolution of human behaviour is best understood as the product of two interacting inheritance systems: genetic inheritance and cultural inheritance.[39] Its advocates see the cultural information shared by a population as analogous to the gene pool of the population. Information in people's heads is passed down the generations and modified over time, like genes. On this basis, they argue, cultural change can be seen as an evolutionary process, a 'descent with modification' that is similar to the evolution of genes. The main difference between genetic and cultural evolution is that culture evolves not just by random processes and natural selection, but also under the influence of human invention and choices. The invention of new ideas and the choices people make about whether or not to adopt existing ideas are major drivers of cultural evolution. Therefore, DIT can be seen as a sort of social constructionist theory.

Dual inheritance theorists argue that the evolution of culture can be analysed using methods similar to those used by population geneticists to analyse genetic change in a population. For example, it is possible to identify evolutionary forces that influence the speed and direction of cultural change.[40] Natural selection is one such force. It influences cultural evolution just as it influences the evolution of genes. For example, people who have better information about how to get food have better survival chances and greater reproductive success. However, more forces operate on cultural evolution than genetic evolution. For example, people often choose, consciously or unconsciously, which cultural variants to adopt. The rules people use that bias their decisions (such as 'do what most people do' or 'do what the most successful people do') act as forces. One of the basic ideas of DIT is that humans have genetically evolved mental characteristics which facilitate the spread and accumulation of useful cultural information.

During human evolution, as our ancestors' reliance on culturally inherited information increased, the process of our species' evolution became different from that of other animals. Genes and culture coevolved; changes in culture influenced genetic change and vice versa. The extreme phenotypic plasticity of humans is one result of this co-evolution. This plasticity brings benefits but also carries risks. An organism with low plasticity is less able to adapt during its lifetime to the specific environment in which it lives. It does, however, have the benefit of a genetic inheritance which makes it well adapted to the average environment in which its recent ancestors lived. Some level of genetic pre-adaptation is obviously necessary and the extent to which phenotypic plasticity is favoured by natural selection is related to the stability of the environment which an organism exploits.[41] Organisms with more plasticity fare better in less stable environments and are better able to colonize new environments. If genetic pre-adaptation is insufficient, however, there is a greater risk that the organism will die before it can develop necessary adaptations.

For humans, flexibility carries relatively low risk because the young develop in a social environment that facilitates their adaptation. Infants and juveniles mature among older conspecifics with experience of living in the environment they share and who have a body of expertise learned from their predecessors.[42] Older conspecifics not only protect and provision the young; they correct their mistakes and encourage them to acquire the necessary knowledge and skills. C. J. Lumsden and E. O. Wilson suggested that humans must possess genetic adaptations that reduce the risk of learning maladaptive behaviours due to carelessness or manipulation on the part of competing conspecifics.[43] Such adaptations might, for example, cause an organism to experience pleasure when considering an adaptive choice and disgust when considering a maladaptive choice. They suggest that such adaptations place limits on human flexibility and keep culture 'on a leash'. While such a leash may exist, it appears to be a very long one. For example, it has long been known that newborn infants prefer sweet tastes to bitter and sour tastes,[44] suggesting that humans have genetically evolved taste preferences. However, the fact that many infants have grown up to prefer double espressos and lemon tea without sugar shows how easily these preferences can be overridden. If an individual can learn from others what is good to eat, genetically evolved preferences are not very useful and could reduce survival by causing individuals to avoid safe and nutritious food that happens to taste a bit bitter or sour. Humans protect themselves from carelessness and manipulation on the part of their teachers by being sceptical and consulting a number of sources. For example, even very young humans do not unquestioningly accept the food that their conspecifics tell them to eat. Children are more likely to develop preferences for foods that their elders consider to be rare treats than food which elders say are 'good for them'. The human strategy of learning what to eat and what to avoid from the people they live with, inevitably leads to the cultural evolution of regional food customs and taboos.

Research suggests that a person's beliefs about sexual attractiveness are also influenced by those he or she lives with. For example, surveys in Zulu communities in Africa show men to prefer a female shape reflecting a much higher body mass index than that preferred by Western men. However, a sample of Zulu males who had emigrated to London were found to have developed a taste for thinner women within a few months of arrival.[45] Humans may lack strong genetically evolved sexual preferences because for most of human evolution, individual preferences played a limited role in their procreative choices. Studies of mating behaviour in the kind of small-scale societies in which most humans lived until about 200 years ago reveal that most reproduction took place within marriage and many areas are so sparsely settled that most people did not have an extensive choice of marriage partner. Meetings between possible partners were organized by networks of kin, and friends and parents exercised considerable influence over whom their children, especially their daughters, would marry.[46] Some evolutionary anthropologists have argued that the custom of arranging marriages is so widespread among hunter-gatherers that it must have originated at a time before anatomically modern humans left Africa more than 50,000 years ago.[47]

We have suggested that the versatility of human sexual preferences can also explain why people engage in same-sex sexuality. But what about modern homosexuality? How can evolutionary theory help us to understand the recent emergence of this new kind of same-sex sexuality? The kin influence hypothesis suggests that it is part of the rapid and dramatic process of cultural change that occurs as new social institutions and new agricultural, industrial and communication technology change the structure of human communities. This process of change, often referred to as 'economic development', changes the composition of people's social networks and the kind of social influence they experience. In communities that are not economically developed, marriage customs and beliefs about what marriage partners should expect from one another vary widely, just as diet and food customs do. And yet anthropologists who have studied these populations from a Darwinian perspective have found an important similarity.

Members of these communities tend to make reproductive decisions which suggest that they are competing to maximize their fitness, just as Darwinian theory predicts.[48] This is not to say that everyone produces children profligately; social norms about who reproduces and when they give birth constrain the rate of reproduction to match the availability of both physical resources and the social resources mothers must draw on for help in raising their offspring. Those who do not reproduce themselves are expected to help their relatives. In this way, families produce the maximum number of surviving offspring possible, given their circumstances.[49]

The European populations, which were the first to undergo economic development, have now experienced over two centuries of rapid, and ongoing, cultural change. Many of these changes have affected reproductive behaviour and caused people to behave in ways that are less likely to result in the achievement of reproductive success.[50] Among the first of the changes that occurred was the adoption of the belief that it is prudent to limit family size and this resulted in a rapid decline in the fertility of the population. The changes that accompanied or followed the fertility decline included changes in parenting norms, gender roles and, of course, the long 'sexual revolution'.[51] Studies of beliefs and values of populations that have undergone economic development more recently have shown that they experience a similar process of cultural change.[52] As more and more of the world's populations have become integrated into the global economy and connected through travel and communication technology, their behaviour is also increasingly diverging from that which is consistent with achieving reproductive success. Fertility is now below two or declining in almost all of the world's populations.

The kin influence hypothesis suggests that economic development triggers the construction of new beliefs about sexuality and reproduction because it creates a social environment in which people form new social groupings and new social identities. Prior to economic development, the family was the dominant social institution for most individuals.[53] People were employed by and educated within their family. They expected to be able to gain help from their extended family if they experienced hard times and they felt obliged to help family members when called upon. Most were illiterate, so social information was communicated through face-to-face interactions among people who lived in the same region.[54] Membership in their family or clan provided an important social identity and maintaining the family was an important goal shared by its members.[55] One would therefore expect these family-based communities to maintain norms which encourage behaviour promoting the welfare and reproductive success of the family as a whole. Such behaviours would tend to maximize what Darwinists call 'inclusive fitness'.[56]

With economic development, individuals become less engaged with their family. They are increasingly employed, educated and entertained by non-family members. They spend more time away from home and become increasingly connected to the wider world through books and other new media. The social psychology that humans have evolved drives them to form new groups. They gain new group identities and their social interactions bring new influences. There is no reason to expect these non-family groups to generate or maintain norms that promote the family or encourage reproductive success, even though behaviour will not change instantly. People will tend to continue to be guided by the norms of the communities where they grew up. Their children, however, are likely to give lower priority to the pursuit of reproductive success and higher priority to interests that compete with reproductive success.

It is noteworthy that new beliefs about, and attitudes towards same-sex sexuality begin to arise following the fertility decline (see Figure 5.1). Norms pertaining to same-sex sexual behaviour vary widely in pre-economic development communities, from outright condemnation to belief that it is a normal part of growing up.[57] However, belief in the existence of a new identity – a new 'species', as Foucault portrayed the modern homosexual – only emerges in societies after they have begun to develop economically.[58] In our view, this new identity owes its existence to a progressive loosening of family-centred reproductive norms, even to the point that modern homosexuals are more or less respected in their decision not to marry someone of the opposite sex and have children.

These new norms and beliefs about family size, love and marriage, and same-sex sexuality, partly create and partly reflect a critical mass of new sexual preferences and relationships. In fact, individuals are challenged to develop sexual preferences to an extent that had not been required in the past. If many generations of arranged marriages had prevented individual sexual preferences from making more than a limited contribution to reproductive success, there is no reason to expect individuals to have inherited genes that ensure they develop sexual preferences that maximize reproductive success. The kin influence hypothesis suggests that human sexuality is quite flexible. Individuals who could learn to accept the partner arranged for them would have had greater reproductive success than those burdened with rigid preferences. However, coupled with an increasing acceptance of same-sex sexual behaviour, and with the new belief that individuals should allow sexual preferences to dictate their choice of sexual and life partner, the flexibility of human sexuality can also result in patterns that are devastating from an evolutionary fitness perspective.

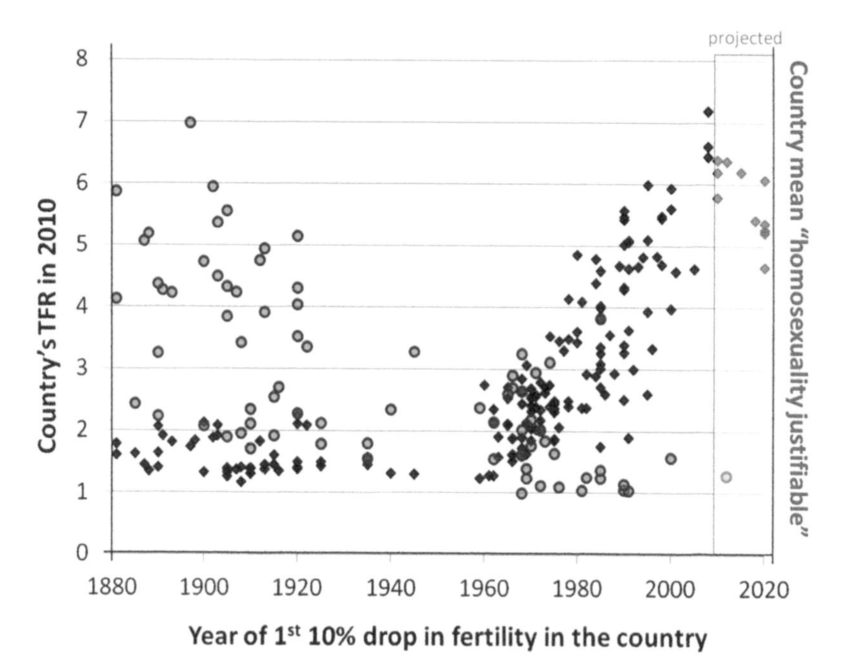

Figure 5.1: Comparison of the total fertility rate (TFR) of countries (black diamonds) with mean attitude to 'homosexuality' in each country (grey circles). The x-axis indicates the year that family limitation began to become widespread in the country or (in the countries in which the fertility has not yet begun to decline) when the United Nations expects it to begin.

The fertility data, compiled by the United Nations (http://esa.un.org/wpp/unpp/panel_indicators.htm), estimate the average number of children born annually per woman. The data on attitudes towards homosexuality were obtained in studies conducted as part of the World Values Survey (see http://www.worldvaluessurvey.org). A representative sample of the population was asked to indicate if they thought homosexuality was justified on a scale from 1 (never) to 10 (always). This measure is imprecise for many reasons. For example, the meaning of words 'homosexuality' and 'justifiable' and their translations undoubtedly varies from country to country. They can therefore only be an indication.

The countries which experienced fertility decline first are European or their population is mostly of European descent. In these countries fertility is currently low, mostly below two children per woman. It is also in these countries where we see greatest acceptance of homosexuality. Some countries, mostly in East Asia, where fertility began to decline in the middle of the twentieth century show moderate levels of acceptance. Most countries where fertility only recently began to decline have very low tolerance of homosexuality.

Assumptions and predictions of the kin influence hypothesis are currently being tested but it is obviously in line with social constructionist claims about modern homosexuality. First of all, it explains why it would be adaptive for humans to have flexible sexual preferences, even though it does not assume that same-sex sexual behaviour itself is adaptive. Secondly, it emphasizes that sexual preferences are profoundly moulded by evolving cultural norms and beliefs. The conflict, confusion and rapid change in beliefs about same-sex sexuality in Western populations in the last 200 years can be seen as a cultural evolutionary process triggered by structural changes in society and driven by the need for the population to reconcile a number of contradictory new beliefs about sexual relationships and a desire to have relatives in future generations. More particularly, the kin influence hypothesis offers a plausible evolutionary account for the emergence of a modern kind of identity-based exclusive homosexuality, and for the virtual absence of such homosexuality before the industrial revolution. Finally, by showing how evolving beliefs have mediated the formation of stable sexual preferences, the kin influence hypothesis suggests that theories, whether scientific or not, have been important in the construction of modern homosexuality – a claim that is central to the social constructionist literature since Foucault.

Still, it remains somewhat unclear why so many theories and beliefs regarding sexual orientation tend to be essentialist. Why do we essentialize this kind of sexual behaviour? In the following section we argue that evolutionary theory can also play a role in answering this question.

Essentializing 'Homosexuality'

Like social constructionism, essentialism comes in many guises. In philosophy, the essentialism debate often revolves around the notion of a natural kind. Some philosophers have argued that natural kinds are such by virtue of their having essences.[59] Natural kind essentialism is true in the case of prototypical natural kinds, such as chemical elements or human blood types. All polonium atoms, for example, are composed of eighty-four protons – a natural feature intrinsic to this chemical element. Having eighty-four protons is the essence of polonium, while the essence of blood type A is the presence of a particular antigenic substance on the surface of red blood cells. Essences of entities not only determine the necessary and sufficient conditions for their being a member of a particular kind; they also explain many other characteristics typical of these entities, such as, for example, their colour and malleability. In short: essences yield a wealth of inductive information about an entity.

However, natural kind essentialism is just one kind of essentialism and philosophers have taken pains to distinguish it from various other kinds. Recently, such distinctions have proven useful in social psychological research. For example: social psychologists N. Haslam and S. R. Levy found that lay people's essentialist beliefs about modern homosexuality come in three kinds.[60] The first set of beliefs is very similar to natural kind essentialism, as it includes the beliefs that modern homosexuality is caused by biological factors, such as genes or hormones, and that it cannot be changed after birth (*immutability factor*). A second set involves beliefs about the historical and cross-cultural invariance of same-sex sexuality (*universality factor*). A final set of essentialist beliefs revolves around 'entitativity' – the belief that 'homosexuality' and 'heterosexuality' can be delineated by means of specific defining non-natural characteristics (*discreteness factor*). Haslam's research does not provide evidence of the relative importance of natural kind essentialism vis-à-vis other kinds of essentialism in conceptualizing sexual orientation; nor did he measure the relative importance of essentialism *tout court*. His research does show, however, that, for lay people, natural kind essentialism about sexual orientation is a very coherent way of thinking, easily activated and readily available.

The question of whether sexual orientations can be considered as natural kinds has exercised many minds since Antiquity. In early modern Europe, for example, physiognomists and astrologists considered various sexual types, including the *kinaidos*, as innately different kinds of individuals.[61] Nineteenth-century psychiatrists and sexologists continued this essentialist tradition with anatomical and degenerationist interpretations, smoothing the way for twentieth-century research in the genetics, endocrinology and neuro-anatomy of sexual orientation.[62] The common denominator in these studies is a desire to delineate the very essence of 'homosexuality'. Modern homosexuals themselves, and particularly gay activists, have also been tempted to appeal to some kind of essentialism to understand and to legitimize their own behaviour and desires.[63] In the second half of the nineteenth century, one of the very first gay activists, Karl Heinrich Ulrichs, argued that specific 'germs' determined a male to be born with a woman's soul, and vice versa. Another German advocate for 'homosexual' rights, Magnus Hirschfeld, believed 'homosexuals' to be neuro-endocrinological hermaphrodites.[64]

In recent years, some philosophers and scientists have argued against an essentialist approach of sexual orientation by claiming that human same-sex sexuality does not meet the criteria needed to qualify as a natural kind. One of these criteria states that, in order for modern homosexuality and heterosexuality to be natural kinds, their members need to be inherently and innately different types of individuals. To ascertain this, one would need a unique and observable set of features, such as genetic or neuroanatomical markers, that sets 'homosexuals' apart from 'heterosexuals'. To this date, scientists have not been able to find such features.[65] And as A. Kinsey already noted in 1948, there are reasonable grounds to be pessimistic about the prospects of their quest:

> Males do not represent two discrete populations, heterosexual and homosexual. The world is not to be divided into sheep and goats. Not all things are black nor all things white. It is a fundamental of taxonomy that nature rarely deals with discrete categories. Only the human mind invents categories and tries to force facts into separated pigeon-holes. The living world is a continuum in each and every one of its aspects.[66]

Another defining characteristic of natural kinds is that they do not bother about the way we classify or conceptualize them.[67] Arguably, chemical elements and blood types do not interact with the concepts and categories with which they are grasped. Modern homosexuals, on the other hand, do seem to be affected by changing conceptualizations of same-sex sexuality. The transformation from same-sex sexuality to 'homosexuality', for example, brought drastic changes to the lives of modern homosexuals. It allowed them to derive their identity from their sexual behaviour, to live together as full and life-long partners and to claim specific civil rights. These changes have, in their turn, adjusted aspects of more recent theories on the topic. In the 1970s, for example, American gay activists were able to overturn the illness theory of homosexuality by showing that many 'homosexuals' live rewarding lives, socially, sexually and professionally.[68] In short, different conceptualizations of same-sex sexuality have differentially affected the lives of modern homosexuals, and vice versa, so it is probably more accurate to consider modern homosexuality as an interactive kind,[69] rather than a natural kind. Interestingly, the interactive kind view does not only contradict essentialist interpretations of sexual orientation, it also nuances early social constructionist views, particularly Foucault's. Contemporary historians of sexuality often fault Foucault for being obsessed with social control, as if labels were unilaterally imposed on individuals by sexologists, psychiatrists or bourgeois society. In their view, the construction of labels and categories, such as 'homosexuality', should be seen as a joint venture between 'patients' and their 'doctors' (a telling example of such interactive constructing can be found in Oosterhuis's book on Krafft-Ebing).[70]

These and other arguments against the natural kind view of same-sex sexuality seem very plausible, and yet the essentialist view continues to be very popular. Why do we feel the need to essentialize sexual orientations? A comprehensive answer to this question is of course beyond the reach of this chapter.[71] It is interesting, however, to note that, in recent years, both evolutionary scientists and social psychologists have suggested that there may be a 'natural' and even 'innate' essentialist bias in categorizing living beings. Susan Gelman found that even young children ascribe natural essences to biological species, but not to artefacts.[72] These 'folk biological' conceptions are obviously at odds with evolutionary biology, as biological species do not have essences in a Darwinian world.[73] Yet essentialism may well be a useful cognitive strategy, because it enables us to make valuable inferences about organisms. The assumption that each and every living kind has a hidden causal essence implies that every new feature it exhibits can automatically be assumed to typify all other members of its species, too. An essentializing mentality, some refer to as a 'folk biology module', thus spares us a costly and laborious learning process.[74]

Surprisingly, many social categories, such as ethnic groups[75] and mental disorders[76] are also frequently essentialized. F. Gil-White has hypothesized that salient similarities between ethnic groups and biological species have led us to use the same cognitive machinery to process information about both entities.[77] It often happens, for that matter, that modules 'erroneously' process information from domains that do not belong to their naturally selected range. Thus it is that most modules have a *proper domain*, i.e. the set of stimuli for which the module has been designed, as well as an *actual domain*, i.e. the set of stimuli triggering the module.[78] Elsewhere, we have argued that mental disorders, just like ethnic groups, are not part of the *proper domain* of what we will refer to as the 'folk biology module', but because of their striking resemblances with biological species, they belong to its *actual domain*.[79] Unlike ethnic groups, however, individuals displaying deviant behaviour, whether sexual or otherwise, did not make a structured population during the bulk of human evolutionary history; and ethnic groups need time to evolve all kinds of salient ethnic markers, such as dress codes and religions, thus creating the illusion that their members are united by some hidden natural essence.[80] So why should we believe that information about mental disorders is processed by the folk biology module?

Answering this question, we have pointed to recent changes in norms and beliefs about mental health. Ever since the dawn of modern psychiatry, various actors and factors have 'conspired' to class patients in supposedly natural and homogeneous groupings. The large-scale institutionalization of mental health care initiated this process by physically segregating the healthy from the ill. In its turn, this institutionalization, often referred to as 'the great confinement', facilitated the construction of blood-based biological theories about mental illness.[81] Such theories, in their turn, precipitated the development of various psychoactive drugs, of which both the intended and the side effects probably contributed to the ongoing process of homogenizing the patient population. All in all, it is easy to see how these and various other elements have fuelled the use of an essentialist mentality – the folk biology module – when thinking about mental disorders.

For a considerable amount of time, psychiatrists also considered 'homosexuality' to be a mental disorder, so perhaps it should not surprise us that sexual differences, alongside ethnic and psychiatric differences, have also been essentialized. In the first two editions of the American Psychiatric Association's *Diagnostic and Statistical Manual of Mental Disorders* (*DSM*), 'homosexuality' was listed as one of the 'sexual deviations', part of the larger diagnostic category of 'personality disorders'.[82] Only in the early 1970s did a new wave of gay activists, with the help of some public intellectuals, succeed in forcing the Association to rethink its classification and to delete homosexuality from the manual.[83] And even though homosexuals have rarely been confined to mental asylums, psychiatrists have taken a very active part in reconceptualizing men who love men to be a type of personality that is inherently different from men who love women. In doing so, they are especially indebted to the work of the late nineteenth-century

Austrian psychiatrist Richard von Krafft-Ebing. Basically, Krafft-Ebing was one of the first psychiatrists to transform homosexuality from a group of behaviours which many people believed to be sinful to a deep and defining characteristic of an individual's personality – a keystone in defining his or her identity.[84] It was Krafft-Ebing that Foucault had in mind when claiming that 'the sodomite had been a temporary aberration; the homosexual was now a species'.[85]

Moreover, the culturally driven formation of homosexual subcultures that occurred in most of the big cities from the sixteenth century onwards also created similarities between the members of these subcultures. Both in eras of prosecution and in relatively more tolerant times, 'homosexual' men had a lot to gain from a well-defined homosexual subculture. At least in large societies, the adoption of a homosexual identity and a homosexual lifestyle is an advantageous strategy, because the social benefits of this identity and especially the benefits of coordinated sexual and non-sexual exchanges probably compensate for the social costs, including the cost of being the target of stigmatization. Furthermore, a well-defined subgroup can be used to manage stigmatization, because it is easier for such a group to claim rights and benefits than it is for a less well-defined group of individuals. As a matter of fact, the architects of the *DSM* admitted that they removed homosexuality from their nomenclature precisely because homosexuals had united themselves in a lobby.[86]

In short, a variety of historical and cultural processes have led to a partly real and partly perceived homogeneity in the population of individuals engaging in same-sex sexuality, thus eliciting the use of a genetically evolved tendency to essentialize observed variation, in thinking about sexual orientation. In casting an evolutionary light on a long-standing issue in the social constructionist literature, i.e., the issue of essentialism, this strand of research can also be seen as an example of evolutionary social constructionism.

Conclusion

Modern homosexuality is an ideal test-case for evolutionary and other naturalistic approaches to social constructions, not in the least because many of the wars between essentialists and constructionists were fought over this very issue. In this chapter, we hope to have shown that adhering to the evolutionary social sciences need not imply an essentialist position. In fact, we believe that a thorough understanding of human evolutionary history, and the pivotal role that culture played in this history, often goes hand in hand with an acceptance of many social constructionist ideas. Much in the same way as Darwin's theory has led to the rejection of species-essentialism in biology; the Darwinizing of culture undermines the plausibility of many essentialist ideas in psychology and other social sciences.

6 DARWINISM AND HOMOSEXUALITY

Gonzalo Munévar

Introduction

Some important counterexamples to Darwin's *On the Origin of Species* were offered by animal behaviour that decrease either the individual's chances for survival or for reproduction. Altruism is an example of the first, homosexuality of the second. Although the problem created by altruism is considered solved by many biologists,[1] the problem created by the existence of homosexuality still awaits solution. Most attempts to solve this problem have tried to show that natural selection and the existence of homosexuality are compatible because homosexuality is in some way adaptive. Against this adaptationist approach I argue that a simpler biological explanation based on gene expression and the variation of traits may suffice to explain the existence of homosexuality. Evolutionary biology, therefore, does not require that homosexuality be adaptive.

It is important to clarify at the start that strict homosexuality seems to be rare: it is limited to humans, domesticated sheep and perhaps a few other unconfirmed species. This does not diminish the problem. According to some claims, over 1,500 species of animals exhibit atypical sex practices,[2] such as same-sex mating, which includes full anal penetration in lions, giraffes, bison and elephants. Gulls form lesbian relationships. Bonobos are a highly bisexual species. And male penguins form long-term homosexual relationships. Many of these animals will often, if given the opportunity, engage also in heterosexual sex, but even so the high incidence of male–male or female–female sex throughout the animal kingdom, i.e. of a multitude of examples of apparently maladaptive behaviour, seems to demand a Darwinian explanation.

Kin Selection and the Explanation of Homosexuality

Given the apparent success of explaining altruism using kin selection by W. D. Hamilton,[3] E. O. Wilson and others have attempted a similar explanation of homosexuality.[4] According to Hamilton, an animal that acts to its own detriment to the benefit of its close relatives increases the chances that its genes will continue to be represented in the population, because its altruistic act increases the fitness of those relatives. Analogously, the homosexuality of an individual would increase the fitness of its close relatives.

Darwin himself had already anticipated kin selection as an explanatory mechanism in evolution.[5] When explaining the existence of sterile females in social insect communities, he wrote:

> This difficulty, though appearing insuperable, is lessened, or, as I believe, disappears, when it is remembered that selection may be applied to the family, as well as to the individual, and may thus gain the desired end. Breeders of cattle wish the flesh and fat to be well marbled together; an animal thus characterized has been slaughtered, but the breeder has gone with confidence to the same stock and has succeeded.[6]

In social insects the family is the community or colony, since in a hive, for example, all the members are related to the queen. For Darwin, thus, 'if ... it had been profitable to the community that a number should have been born capable of work, but incapable of procreation, I can see no special difficulty to in this having been affected through natural selection'.[7] The reason is that the high efficiency that results from specialization has been advantageous, and, since selection has applied to the family, 'the fertile males and females have flourished, and transmitted to their fertile offspring a tendency to produce sterile members with the same modifications'.[8]

Just as Darwin uses division of labour to account for the sterility of the workers in social insect communities, Wilson has tried to use division of labour also to account for the existence of homosexuality.

In Wilson's account, the existence of homosexuality in humans is explained by hypothesizing that in hunter-gatherer populations, a homosexual individual, not saddled by the drive to procreate, could more easily achieve high status within the group. By becoming a shaman, for example, a homosexual man could make extra resources available to his nephews and nieces, thus directly enhancing their chances of survival and procreation, and indirectly enhancing the chances of the genes he shares with them to be passed on, including presumably those that make him homosexual.[9]

Unfortunately, Wilson's explanation may strike many readers as an implausible just-so-story, a paradigm example of what Stephen J. Gould used to call 'adaptationism', i.e. the notion that every trait exhibited by an organism must be the result of an adaptation.[10] In any event, however interesting we may find Wilson's speculations, we will see below that the solution to the problem need not lie along this path. In particular, we will see that the existence of many developmental stages between genome and phenotype should make us weary of accepting an account such as Wilson's without some additional evidence, which no one has produced, as far as I know.[11]

There have been many other approaches to reconciling natural selection and the existence of homosexuality. I will mention two here that strike me as particularly interesting:

(1) An analogy to the explanation of the existence of sickle-cell anaemia. The heterozygous stage produces such great advantages (resistance to malaria) that it makes up for the fatal condition in the homozygous state. Similarly, the gene that 'causes' homosexuality in the homozygous state produces some great advantages in the heterozygous state.[12] In less technical terms, a recessive gene present in either the mother or the father would convey some advantages, but when the gene is present in both parents, the offspring will be a homosexual (or at least a bisexual). Unfortunately, so far those advantages are completely unknown, and the relevant gene or genes are equally unknown.

(2) According to a study by A. Camperio-Ciani et al., the female relatives of homosexual men, on the mother's side, have more offspring than those of heterosexual men.[13] Presumably the same gene makes both females and males strongly attracted to men. It would remain in the population because the strong reproductive success of the females makes up for the lower reproductive success of the males. Nevertheless, no one has identified any such 'homosexual genes', and thus explanations of this type remain highly speculative.

This is not to suggest that there has been a paucity of attempts to establish the heredity of homosexuality and to identify the 'homosexual genes' that would clinch the matter. Indeed, we can find in the literature several references to sibling studies, twin studies, candidate-gene studies and linkage studies. R. C. Pillard and J. D. Weinrich, for example, report that the brothers of homosexual men have a 22 per cent chance of being homosexual.[14] J. M. Bailey and D. S. Benishay also reported a similar probability for the sisters of lesbians.[15] Simon LeVay points out that although these results are consistent with genetic influence, they do not 'distinguish between genetic and environmental causes'.[16] Similarly, twin studies depend on the concordance rate for homosexuality between twins, which is high for all twins but particularly for identical twins (22 per cent to 52 per cent in males, and 16 per cent to 48 per cent in females, according to some studies).[17] LeVay acknowledges that such studies are highly suggestive but points out that, like other sibling studies, they do not quite separate genetic from environmental factors (e.g., identical twins being treated very similarly), which makes it difficult to determine heritability in terms of concordance rates.[18] J. P. Macke et al. hypothesized that the androgen receptor gene (important in testosterone) might influence sexual orientation. Their study, though, found no significant differences between homosexual and heterosexual men.[19] D. H. Hamer et al., on the other hand, claimed to have found in the X chromosome a gene likely to influence male homosexuality (in region q28).[20] Independent attempts to replicate their findings, however, have been unsuccessful.[21]

In spite of the shortcomings of these attempts, their results are encouraging enough to justify further investigation. Success in such endeavours would immediately prompt the need to explain how such genes remain in the population in spite of their seemingly maladaptive character. Nevertheless, whether such future studies are ever successful or not, my contention is that Darwinism is not inconsistent with the bisexual and homosexual behaviour exhibited by animals.

A Different Approach to Account for Homosexuality

Although a *biological* explanation of the existence of homosexuality is still called for, it need not depend on its resulting from the action of natural selection. A perhaps more reasonable candidate may be found by considering anew Darwin's thoughts concerning the principle of variation of characteristics. 'Some authors use the term "variation" in a technical sense', Darwin wrote, 'as implying a modification directly due to the physical conditions of life; and "variations" in this sense are supposed not to be inherited.'[22] Nevertheless, he pointed out, 'When a variation is of the slightest use to any being, we cannot tell how much to attribute to the accumulative action of natural selection, and how much to the definite action of the conditions of life.'[23] A similar point may apply, I presume, to variations that are disadvantageous. Our problem is the converse of Darwin's, who was trying to carve out a role for inherited variations on which natural selection could act. If such disadvantageous variations, such as homosexuality, were not at all due to the 'action of the conditions of life', then natural selection would act to eliminate them. My suggestion is, then, that variations in sexual preference are indeed subject to the action of the physical and even the social conditions of life. As I will explain below, this result would make bisexuality in those 1,500 species, and homosexuality in humans and rams, an expected result from their biological nature.

Consider whether sexual preference is likely to be distributed in an animal population as so many other phenotypic traits are, say the width of noses (see Figure 6.1). In the case of the width of noses, Darwin could point to the fact that children are likely to resemble their parents, and, thus, even though some amount of variation in the width of the nose should probably be attributed to the action of diet, or the degree of climatic temperature and humidity in the early life of the person, for the most part we can be quite confident of the heritability of the trait.

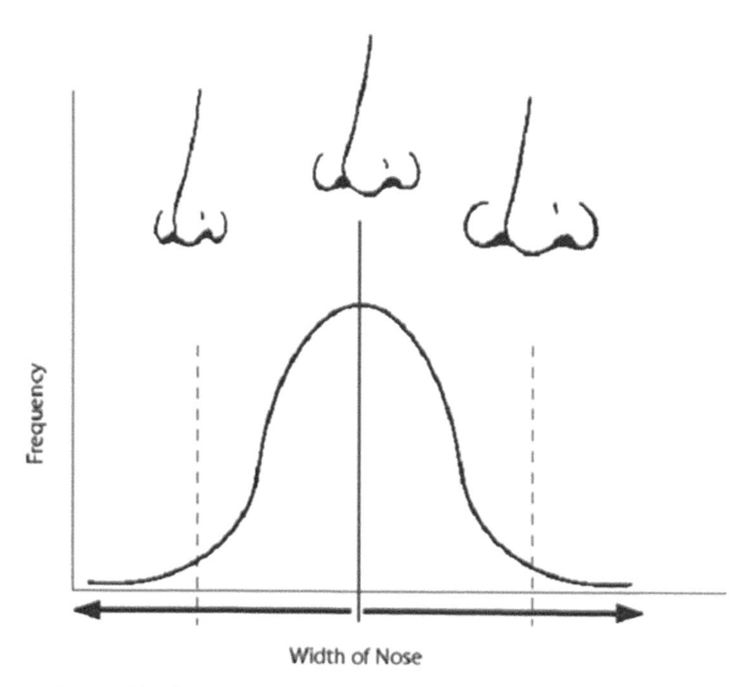

Figure 6.1: Width of nose distribution. Illustration by Nicole Ankeny. When observing many biological (in this case, phenotypic) traits in a large population we get a bell-shaped distribution. Human noses are a good example: we see narrow noses and wide noses, with the majority of noses falling within two standard deviations of the mean average.

On the other hand, the case of sexual preferences is different. In complex animals, according to the work of LeVay and others, several stages of development have to work in specific ways for the individual organism to have a clear preference for the opposite sex. Thus, it seems to me, the genes responsible for making males be attracted to females and vice versa will have to interact with, say, the embryonic environment, and in some cases with the social environment also, as we will see below. Those interactions, furthermore, will be subject to 'noise', e.g. disturbances in the chemical medium. As is well known, however, genes that interact with the environment and with noise yield a continuous distribution curve in the expression of the trait.[24]

LeVay also points out a contrast between the expression of the genes for sexual preference in insects and in mammals. In Drosophila, for example, a single gene named *fruitless* ('fru') is read off the genome into messenger RNA, and thus leads to different proteins, in the neurons of male and female fruit flies. These neurons are probably involved in the detection of sex pheromones. The cells can then be said to be autonomous in determining sex differentiation in drosophila and other insects, whereas in humans and other mammals such differentiation 'depends in large part on circulating sex hormones'.[25]

Of course, genes do influence sexual preference.[26] A gene situated in the Y chromosome, called 'testis-determining factor' or TDF for short, develops gonads into testis. Since females have no Y chromosome, TDF is absent and the gonads develop instead into ovaries. The testis and ovaries then influence other sex differences in the body, such as internal and external genitalia as well as the brain. In males, for example, TDF also leads to the development of Sertoli cells, which in puberty in turn lead to the development of sperm, and while still in the embryonic stage instruct some other cells into becoming Leydig cells, which among other things secrete testosterone and other androgens, steroid hormones that help bring about structures and behaviours typical of the male individual. If this rather long process is not present (no TDF) the body will develop into a female, the default setting, one might say, of sexual preference.[27]

Once begun, the concentration of testosterone rises and falls, affecting the development of a variety of structures in the body, including the brain. It rises at fourteen weeks after gestation, two months after birth, and during puberty. Although 'programmed', these variations on concentration may be affected by the internal or external environments, as well as by 'noise', in turn thus affecting, for example, the development or wiring of brain structures that play a role in sexual preference (see Figure 6.2). That is, as I will illustrate in what follows, the behavioural traits that result from the relevant genes will tend to exhibit a smooth distribution, from very strong to very weak heterosexuality.

In the normal development of the external genitalia, for instance, a converting enzyme (S-alpha-reductase) converts testosterone into another steroid called 'dihydrotestorone'. If the converting enzyme is not present in sufficient levels, the external genitalia will not be completed as that of a male. The proper level of testosterone must then be present at the right time in the right environment (see Figure 6.2). In the brain, some cells use aromatase to convert testosterone into an important estrogen called estradiol (males require some low levels of female sexual steroids, just as females require some levels of testosterone). This process is important for the programming during prenatal life and infancy of male sexual behaviour, as indicated by studies on mammals.[28] External environmental factors that cause stress in pregnant rats adversely affect the production of estradiol and reduce the size of a hypothalamic nucleus that is key to male heterosexual behaviour (to be discussed below).[29] The male offspring of those rats exhibit less masculine behaviour in later life.

The balance of sex hormones that guides development is complex and subtle. Having too much of a hormone may lead to atypical sex behaviour. Excessive amounts of androgens produced in a condition called 'congenital adrenal hyperplasia' in female foetuses will cause later an increase in male sexual characteristics. The variety of factors that affect the expression of genes involved in sexual preference should lead us to expect, once again, a distribution of phenotypic traits. Moreover, accidental circumstances, which from a perspective that sees genes as containing instructions must be considered noise, also play a part in the expression of those genes. For example, female rat foetuses that lie in their mother's uterus directly downstream from males are more likely to mount other females later in life.[30] Position in the placenta, though, is purely a matter of accident.

The combined action of this great variety of factors is a great variability in sexual behaviour, including behaviour directed to members of the same sex. Thus, as LeVay points out, 'Female rats vary in the frequency of lordosis and in circumstances in which it can be elicited' whereas, as we have seen, 'it is not uncommon for female rats to mount other receptive females, to perform pelvic thrusting and even to show the motor patterns that accompany ejaculation'.[31] Furthermore, 'Diversity in sexual behavior is found among male rats too.'[32]

The rats that are more active when paired with receptive females have a larger sexually diamorphic nucleus in the hypothalamus than the less active rats.[33] In a famous study, LeVay discovered that in human heterosexual men the volume of that very nucleus is about three times larger than in females, *and about two to three times larger than in homosexual men.*[34] As we can see in Figure 6.3, the nucleus, the third interstetial nucleus of the anterior hypothalamus (INAH3) is found in the medial preoptic area of the hypothalamus.[35] To place this finding in the context of the previous discussion about hormonal action, development and the diversity of sexual traits, I would like to refer to the work of Anne Perkins and her colleagues, who found that in the medial preoptic area of heterosexual rams the level of aromatase is about twice as high as that of homosexual rams.[36] As it turns out, INAH3 is also larger in heterosexual rams than in ewes or homosexual rams.[37]

In heterosexual males, both in humans and rams, the INAH3, in the medial preoptic area, is much larger than in females and than in homosexual males. In females, the ventromedial nucleus offers an equivalent anatomical difference.

LeVay was highly criticized because in his study he used the brains of dead HIV patients. Many critics felt that the HIV might have been responsible for the difference in size of the INAH3, and others, particularly W. Byne and B. Parsons,[38] were initially unable to replicate his results when controlling for HIV. Nevertheless, a subsequent study by Byne himself yielded results that LaVey feels are consistent with his own. Indeed, in Byne's own words: 'HIV status significantly influenced the volume of *INAH1* (8 per cent larger in HIV1 heterosexual men and women relative to HIV2 individuals), *but no other INAH*.' Thus, the presence of HIV did *not* influence the size of INAH3. Moreover, Byne did find a difference in size between the INAH3 nuclei of homosexual and heterosexual men. Again, in his own words: 'Although *there was a trend for INAH3 to occupy a smaller volume in homosexual men than in heterosexual men*, there was no difference in the number of neurons within the nucleus based on sexual orientation.'[39] Part of the criticism against LeVay is that he should have considered not the size of INAH3 but the number of neurons in them. And indeed apparently Byne found that the neurons in homosexual men (in INAH3) are more densely packed.

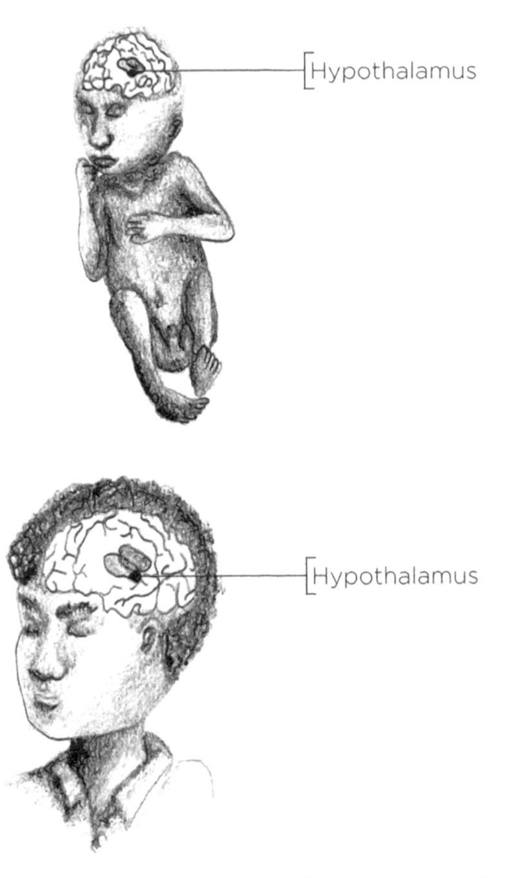

Figure 6.2: Development of the hypothalamus. Illustration by Nicole Tischler. Neural circuits that contribute to male orientation and are wired in the fetus' hypothalamus become activated by testosterone during puberty.

This suggests, to me, fewer dendritic connections, and thus functional differences. There may also be a greater number of improper connections, simply because the normal pattern of neuron development is altered. In a similar case, analogous problems have been found in the hippocampus of schizophrenics, which is smaller than the average hippocampus.

It may be that LeVay will turn out to be mistaken on his results concerning the INAH3, but in the meantime the concurring evidence from comparative neuroanatomy, particularly rams, is very suggestive.

Another relevant finding concerns the amygdale of sheep, a brain structure that is also sexually dimorphic in humans, so that the right amygdala has more connections than the left in heterosexual men, whereas a heterosexual woman's left amydgala has more connectivity than the right.[40] In homosexual men and women those patterns of neuronal connectivity are reversed. Interestingly enough, Perkins and her group found that in the amygdala of heterosexual rams the level of estrogen receptors is about four times higher than in that of the ewes or the homosexual rams.[41]

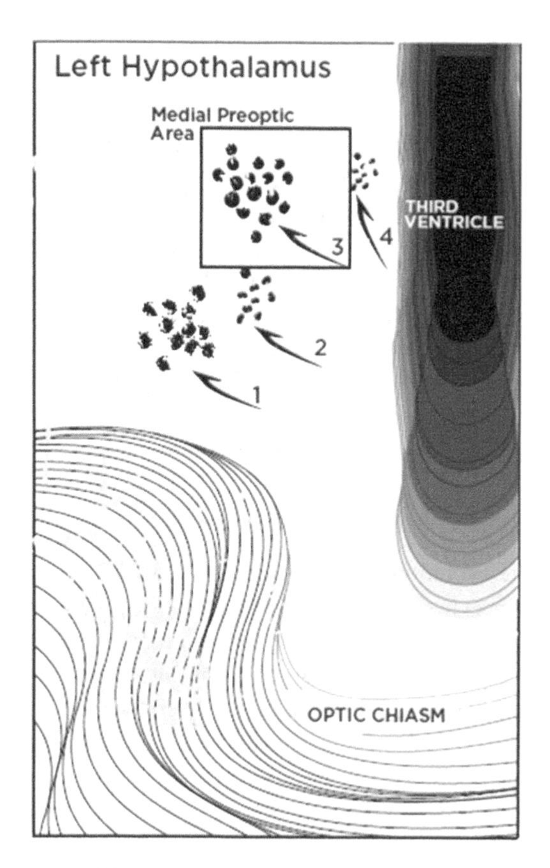

Figure 6.3: Image of the third interstitial nucleus of the anterior hypothalamus (INAH3). In heterosexual males, the INAH3, in the medial preoptic area, is much larger than in females and than in homosexual males. Illustration by Nicole Tischler.

Apart from the variation of the brain and other structures through which the genes relevant to sexual preference are expressed, the timing of the processes that trigger such expression is sure to add even more variation to sexual behaviour. If the synchronization of testosterone and some particular enzyme is off because either arrives late, the structure in question (e.g. INAH3) may be only partially realized. Those triggers, incidentally, are not limited to the prenatal environment, as the example of puberty in humans illustrates quite well. I will offer two additional illustrations.

After birth, the mother rat will lick the anogenital areas of her pups, but will lick the males longer, guided by their scent. If her olfactory nerve is severed, she will lick all pups equally, with the result that the males will be less likely to mount females as adults.[42] Apparently the mother's licking of the anogenital area stimulates the production of testosterone in the male pups, which at that particular stage of development is needed if they are ever to fully express their genetic potential regarding sexual preference.[43]

Rhesus monkeys raised in isolation from peers their age also fail to mount females as adults. The reason is that in play-acting with other young male monkeys they practice, among other things, the double foot-clasp mount characteristic of their species. Without this social stage, the appropriate genes will not be expressed.[44]

Humans, with our greater degree of neoteny and social complexity, are likely to have the development of our sexual preference influenced by a great many subtle and complex factors. We are, therefore, likely to exhibit a distribution of sexual preference. That is, we should expect the appearance of human bisexuality and homosexuality on purely biological grounds (i.e. given how our genes for sexual preference should be expected to express themselves), but without appeals to homosexuality as an adaptation.

This point is strengthened by the following considerations.

The Complexity of the Erotic: Non-Procreational Functions of Sex

According to the evolutionary explanation of altruism mentioned earlier, a mother who sacrifices herself for her offspring acts so as to preserve her inclusive fitness. But she is not motivated by the calculation that by saving her children she will have her genes passed on to later generations. Love is her motivation. In beings with complex brains, natural selection has brought about indirect mechanisms that motivate the adaptive behaviour. Pleasure, for example, rewards the satisfaction of the desire for sex. But even if such desire, combined with the standard attraction to the opposite sex, increases the chances for procreation, it can be satisfied in a variety of ways. Animals may then come to seek sexual pleasure for its own sake, even when the evolutionary 'justification' (procreation) is not an option, e.g. in masturbation.

Abilities evolved to serve specific adaptive goals can be, and often are, used for a variety of purposes. Many skills that were selected because they served our hunter-gatherer ancestors well are now called into service to play a game of soccer or dance a fiery tango. It is not surprising, then, that sex, including the act of copulation, should have been co-opted for purely recreational or social activities by a great many animals. In giraffes, for instance, the percentage of male-male mounting has been reported to be as high as 94 per cent of all mountings.[45] This high incidence of same-sex behaviour could serve to reduce intra-species aggression, exert dominance, or facilitate bonding, as presumably happens in other species. In bonobos, for example, females resolve conflicts and socialize new members into the group by frequently engaging in genito-genital rubbing.[46] Indeed, sexual contact between bonobos is common (about 60 per cent between females) and it seems to occur in most situations in which conflict could otherwise arise, or has arisen, between two individuals.

These examples illustrate once again how the indirect expression of the relevant genes may give rise to great variation in sexual behaviour. Humans, of course, are not social in the way that giraffes and bonobos are, but given the complex development and interconnectivity of our brains, it should be expected that several of the mechanisms that lead to sexual preference will influence, and will be influenced by, many of our experiences in our own prenatal and postnatal histories. That is, what we find sexually stimulating or erotic will itself be complex, thus leading to great variation in the sexual behaviour of our species.[47]

Exclusive Homosexuality and a Prediction of the Non-Adaptionist Model

One puzzle for any account of sexual preference is the existence of strict or exclusive homosexuality, that is, the rigid preference for partners of the same sex even when members of the opposite sex are available. We find it only in humans, rams, and perhaps a few other species. One plausible explanation, however, can be found in the development of INAH3 and other allied structures, which is not the same in all animals. Whereas in male rats lesions to the medial preoptic area leave them sexually interested in females, although confused as to what to do with them, similar lesions in primates make them lose all interest in females, even though their ability to masturbate shows that they continue to have a sex drive.[48] Although we have most genes in common with our fellow primates, particularly chimpanzees, we differ from them in the way some of our genes regulate the expression of other genes. But development may affect the expression of such regulation also. If our developmental mechanisms for sexual preference are par for the course, those differences in regulation (with respect to other primates) will be present in them as well. It may then be, for example, that a significantly small INAH3 and a large or active ventromedial nucleus will be coupled with failure to achieve some other stages of masculinizing development, and all these factors in combination would strictly limit the source of sexual stimulation to other men.[49] A similar explanation would probably be found for sheep. And perhaps such explanations would give hints as to how to find a corresponding explanation for exclusive female homosexuality.

These considerations lead to a prediction that seems to come quite naturally from the non-adaptationist model of explanation I have employed. If becoming female is the default mode for human beings, and a much greater number of processes regulated by hormones, and so on, must take place for a human being to become a heterosexual male, there is a greater chance of failure in the achievement of male than of female heterosexuality in humans. My review of the most significant studies of the prevalence of male and female homosexuality, ranging from occasional to exclusive homosexuality, indicates that the prediction is true. Beginning with Kinsey's studies in 1948 and 1953,[50] in which he found rates of strict male homosexuality around 3–4 per cent and around 1–2 per cent for female homosexuality, to re-evaluations of his data in the 1970s, to more recent studies that have lowered the absolute numbers (around 2 per cent and 1 per cent respectively),[51] the ratio remains pretty well constant across studies: There are roughly about twice as many male as female homosexuals. I suspect the field may need greater methodological rigor, but the results are quite suggestive nonetheless.

Darwin's Principle of Variation Revisited

We have seen that a great many factors may affect the expression of the genes that play a role in sexual preference. The resulting phenotypic variation does not preclude the existence of genes that actually predispose the carrier to homosexuality. Their existence might be masked from natural selection because, even if deleterious, they affect or are affected by other genes that have more beneficial effects, where the adaptive value of the ensemble is positive, or at least neutral.[52] Or the trait in question, e.g. homosexuality, may be polygenetic, i.e. controlled by many genes, particularly when interaction with the environment leads to peculiar combinations of the actions of those genes. For example, the epigenetic markers on one particular gene in one particular environment may lead to a slight alteration in the combined action of other genes it regulates, leading, say, to a lower production of aromatase in the medial preoptic area of the hypothalamus.[53] In any event, an ensemble of co-regulating genes may be on occasion maladaptive even though in most likely developmental scenarios, i.e. on balance, it would be advantageous.

The account defended so far in this paper need not preclude the possibility that some homosexuality genes might follow the model of sickle-cell anemia, nor that, alternatively, some male homosexuality genes make the female relatives hyper-heterosexual, thus leading to compensatory fecundity on their part. If such possibilities came to be realized, they would add to factors that lead to a distribution of sexual preference traits, for clearly no such new factors would prevent the operation of those others I have presented as leading to that distribution.

Indeed, it may even turn out that some mutations do incline the individual to exclusive homosexuality and that the consequent reduced fertility would lower the frequency of such genes in the population and would eventually eliminate them.[54] That situation would return us to Darwin's complaint about variation, namely that 'we cannot tell how much to attribute to the accumulative action of natural selection, and how much to the definite action of the conditions of life',[55] when the "conditions of life" are precisely the many factors that lead to a distribution of sexual-preference traits. Given the evidence presented so far in this essay, it would seem likely that the overall reduction in the rate of homosexuality would be small and difficult to detect, thus making Darwin's task particularly onerous. Of course, such a turn of events would not undermine the non-adaptationist model of explanation I have defended.

Conclusion

From all these considerations it seems reasonable to conclude that explaining the existence of bisexuality and homosexuality does not require showing that either is adaptive. Other biological considerations, particularly about development, lead us to expect a distribution of traits. In humans, we would probably find hyper-heterosexuality at one end, a majority preference for the opposite sex around the mean, and exclusive homosexuality at the other end. As with other natural sciences, the solution to a problem in evolutionary biology may be found in related but different fields, in this case: in an analysis of the ways in which the genes relevant to sexual preference may be expressed.

Acknowledgements

I wish to thank Julie Zwiesler-Vollick, Hsiao-Ping Moore, David Paulsen, Peter Hadreas, Inmaculada de Melo-Martín, Filomena de Sousa and Shelton Hendricks for their comments. I also wish to thank Nicole Ankeny and Nicole Tischler for their illustrations, and my research assistant Yi Zheng for all her editorial help.

7 THE EVOLUTION OF FEMALE ORGASM: NEW EVIDENCE AND RESPONSE TO FEMINIST CRITIQUES

Elisabeth A. Lloyd

Introduction

Let us consider adaptive explanations from evolutionary theory. Take the timber wolf, one of Darwin's examples. Descended from more generalized and slower carnivores, the wolf evolved specialized traits for hunting swift prey like deer and elk. For any explanation by natural selection, we need to specify not only the traits, but the hereditary basis, connection to fitness and the environmental pressures, as sketched below for the wolf case.[1]

Wolf Selection Model

Population: wolf population
Trait: speed: metatarsal to femur ratio
Hereditary basis: genetic basis for ratio
Connection to fitness: speedier wolves catch more prey
Selection pressure: fast prey elude hungry wolves

We start with variation in the crucial traits of swiftness, slimness and strength, Darwin writes, and because there was a reproductive advantage associated with these traits, we have the wolf's specialized adaptations for swiftness that contribute to their reproductive success (or fitness) today. What happens in the population as a whole is that the mean speed of the wolves, which is highly variable in the ancestral population, increases over evolutionary time, as the trait is selected generation after generation, until there is a peak in the population distribution of speed, at a high speed.

The Obvious Account

When thinking about female orgasm, at first it seems perfectly obvious that the trait is an evolutionary adaptation: women who have orgasms will want to have more intercourse, and more intercourse leads to more babies. Thus, female orgasm seems to be an adaptation evolved through natural selection, which explains its presence and spread in the human population. Unfortunately, the basic premise of this explanation is not true. Women who have more orgasms do not have more intercourse, as Alfred Kinsey and his colleagues established in the 1950s.[2] It may be that women who experience more sexual excitement experience more intercourse, but sexual excitement is not the same thing as orgasm, as any woman can tell you. More importantly, women who have more orgasms do not seem to have more babies. In fact, there is no known association between the ability of women to have orgasms and their success in leaving offspring, thus there is no known tie between evolutionary fitness and female orgasm. Also, if the obvious explanation were true, would we not expect to find that the human population would be full of women who were fully and easily orgasmic with intercourse? Yet only about 10–15 per cent of women reliably have orgasm with intercourse, as established in both recent studies and across a span of eighty years of sex research.[3] So the simple and obvious explanation simply conflicts with the evidence.

The evolution of female orgasm is an important puzzle to work out, with real consequences. For instance, in medicine, an evolutionary account of female orgasm is used to underpin a notion of the normal function of female orgasm for the diagnostic and statistical manual of the psychiatric profession, the DSM 4, by which sexual disorders are diagnosed.[4] The business of deciding whether or not women are 'normal' based on their orgasmic performance according to some adaptationist accounts has, arguably, been very destructive, since it has held them to an unreasonable standard. But in evolutionary science, there is no single accepted evolutionary account of female orgasm, and thus no notion of its normal function, if it has any at all.

Pair-Bond Explanations

Now let us examine a very popular alternative adaptive explanation, one that serves as the basis of ten of the available twenty-one evolutionary explanations for female orgasm.[5] It is based on the notion that a male–female pair bond is adaptive, and that female orgasm helped with pair bonding.[6] A pair bond is an enduring attachment, usually monogamous, between a particular man and woman. The pair bond is supposed to be adaptive for several reasons, especially in that it helps with the raising of the very dependent human infants. There have been many feminist analyses and critiques of the pair bond explanations, as they often resemble female stereotypes rejected by most feminists. As animal behaviourist and human evolutionist Desmond Morris claimed, 'the vast bulk of copulation in our species is obviously concerned not with producing offspring, but with cementing the pair-bond by providing mutual rewards for the sexual partners'.[7]

Morris is one of the leading pair bond theorists. Even though his book, *The Naked Ape* (1967), was a popular book, his work greatly influenced other scientists, who adopted his pair bond view.[8] More specifically, the pair bond was taken to be adaptive, and frequent intercourse was believed to 'cement' the pair bond (in Morris's words), while female orgasm served as the steady reward for the female to participate in frequent intercourse.

These relations between traits are represented in this adaptive pair bond model.

Pair Bond Adaptive Model:

Trait: orgasm with intercourse
Trait: frequent intercourse (tied to orgasm rate)
Trait: strength of pair bond (tied to frequent intercourse)
Hereditary basis: genetic basis of female orgasm
Connection to fitness: pair bond increases care of children
Selection pressure: hostile environment for provisioning and protecting children

Morris adopted the Masters and Johnson model of sex, under which the clitoris receives indirect stimulation from intercourse, but on Morris's account, such stimulation leads almost *always* to orgasm for the woman.

However there is a problem with this explanation. Women's orgasm does not occur almost always with intercourse, as these pair-bond theorists assume. Let us return to the wolf and its evolution of speediness. Recall that average wolf speed is very high, after selection over evolutionary time, and that the population is concentrated at this peak speed. Now compare this to the trait of orgasm with intercourse among women.

In Figure 7.1, note that only about 13 per cent always have orgasm from intercourse, and over a third of women rarely or never have orgasm with intercourse, while at least 10 per cent of women never have orgasm from any means. The rest fall in between, sometimes having one and sometimes not. Note also that the Figure 7.1 curve is basically flat. Simple directional or balancing selective forces usually produce *peaks* in the distribution curves of a trait. (There are exceptions to this expectation; for example, one form of sexual selection involving female choice may produce increased variability in the trait, which I address below.) The fact that women vary so widely in their orgasmic performance thus provides very suggestive evidence that twenty of the twenty-one selection explanations offered so far for female orgasm – no matter what their details are – are wrong, since they offer simple directional selective forces. Each one of these explanations, including the pair-bond explanation, predicts a peak in this curve, like the wolf curve; but there are no peaks. And the low rates of women's orgasm with intercourse are not compatible with these evolutionary pair-bond accounts that claim a constant female orgasmic reward for intercourse.

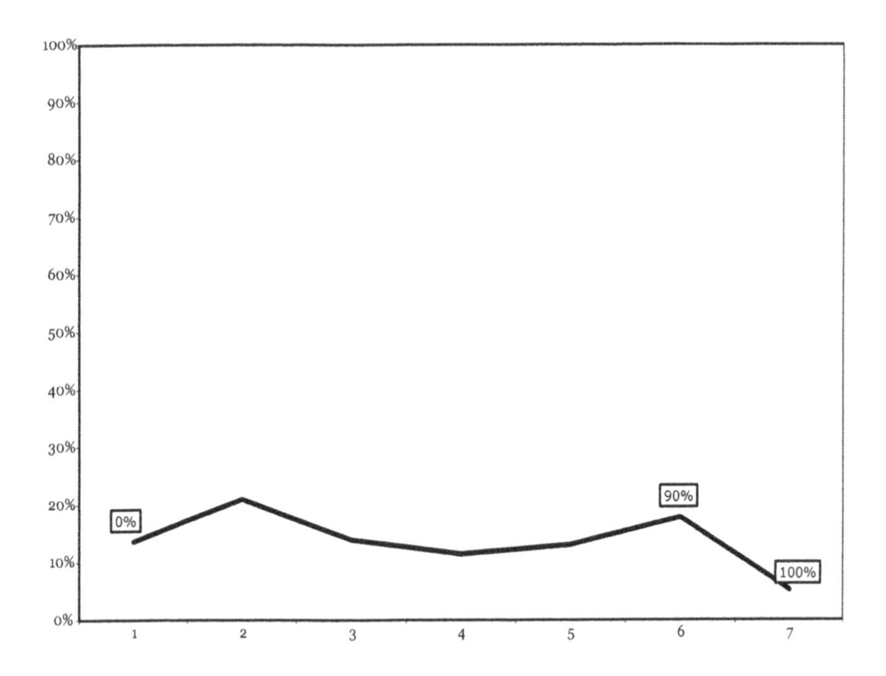

Figure 7.1: Women's orgasm rate with intercourse: x-axis of overall orgasmic performance by percentage; y-axis represents frequency in the population.[9]

A further problem with the pair-bond accounts, which assume that females engage in intercourse for the reward of orgasm, is that women who are more orgasmic do not have more desire to have sex. They do not have higher libidos, and women who do not have orgasms do not have lower libidos. There is no correlation between how orgasmic a woman is and her having – or being interested in having – intercourse. This was first established by Kinsey and colleagues, but has been confirmed by a new study of nearly 3,000 women.[10]

Here is the status of the evidence for the pair bond account:[11]

Pair-Bond Adaptive Model:

Poor	Trait: orgasm with intercourse
None	Trait: frequent intercourse (tied to orgasm rate)

Unknown	Trait: strength of pair bond (tied to frequent intercourse)
Good	Hereditary basis: genetic basis of female orgasm
Fair	Connection to fitness: pair bond increases care of children
Fair	Selection pressure: hostile environment for provisioning and protecting children

Thus, there are grave evidential problems with the ten proposed pair-bond accounts – they're not compatible with the sex research. So what drove scientists to these flawed explanations? My conclusion is that there is a male-centred bias operating, for throughout the explanations, female response to intercourse is *assumed* to be identical to male response to intercourse. This unfounded and unconscious assumption has led many researchers to pursue misguided theories. Kinsey and colleagues showed that males' response to intercourse is almost always orgasm. And in these pair-bond explanations, female response to intercourse is assumed to be *identical* – that is, orgasm – as a reward for intercourse.

Male-centred bias involves looking at things from an exclusively male point of view, and subsequently neglecting a distinct treatment of a female point of view.[12] In these scenarios, an autonomous or distinctly different female sexuality, which emerges so clearly from the sex research, is ignored. And this is despite the fact that several of these evolutionary accounts actually *cite* the sex research containing the crucial information that undermines their accounts.

Male-centred bias in this case was resisted by anthropologist Sarah Blaffer Hrdy, who offered a female-centred explanation based on females soliciting multiple matings from potentially infanticidal or aggressive males.[13] Unfortunately, that explanation also conflicts with the sexology evidence, which I reviewed in my book *The Case of the Female Orgasm* (2005) and I shall not repeat that discussion here.[14]

Oxytocin

Some readers may be wondering about oxytocin, the neurohormone associated with pair-bonding, love, nurturing behaviour and also orgasm.[15] It has been suggested that oxytocin may be a way to rescue a pair bond account of female orgasm, or to provide an independent one, and I recommended pursuing this line of research *The Case of the Female Orgasm*. Experiments have shown that injections of oxytocin can cause uterine contractions that lead to the rapid transport of material near the cervix up into the upper reproductive tract.[16] As orgasm has been documented to release oxytocin, it is hypothesized that orgasm would thus help move sperm up the tract.[17] But this notion that the orgasm actually selectively increases fertility through oxytocin release has recently been debunked by physiologist Roy Levin, who pointed out that the levels of oxytocin used in the sperm transport experiments were approximately 400 times a woman's natural levels of release with orgasm.[18] This does not mean that the phenomenon does not happen, but makes it much less likely that it makes a difference to fertility. Remember that there is no known correlation between orgasm and reproductive success, which is compatible with the lack of effectiveness of any such sperm transport by oxytocin.

Female Choice Explanation

Recently, most of those pursuing an adaptive account of female orgasm have relied on a female choice sexual selection theory of the trait.[19] As mentioned before, it is the only adaptive account currently proposed that is compatible with the flat curve distribution in Figure 7.1. The basic idea is that the female will mate with more than one male over a short period of time, and have orgasm preferentially with the higher quality males. (Higher quality males are recognized unconsciously by females by such traits as higher symmetry values.)[20] The assumption here is that orgasm is accompanied by a mechanism of uterine upsuck (like the oxytocin mechanism) that makes it more likely that the female will be fertilized by the higher quality male.[21] Thus, orgasmic women are required to respond with orgasms only sometimes with intercourse: 'yes' with high quality males, and 'no' with lower quality males. Here is a sketch of the female choice selection model, along with estimations of the state of the empirical evidence supporting the various aspects of the model.

Evidence for the Female Choice Adaptive Model:

Poor Trait: orgasm preferentially with high quality males
None Trait: uterine upsuck
Fair Trait: multiple matings within short period
Poor Trait: conditionality of orgasmic response
Good Genetic basis: orgasm's heritability
None Connection to fitness: through increased fertility with high quality males
None Selection pressure: on women to have offspring of high quality fathers

I would like to draw your attention to two aspects of this adaptive model. First, while uterine upsuck is commonly assumed by the female choice model, it has recently been declared utterly devoid of evidence.[22] The hypothesis had achieved widespread acceptance since the nineties through the work of Robin Baker and Mark Bellis,[23] which was claimed to provide empirical evidence supporting the phenomenon. But consider their data. In one key data set, they have one out of eleven couples in the sample contributing ninety-three out of the 127 data points (nearly three quarters of the data). Four of the other ten couples contributed one data point each, a combined total of 3 per cent of the data, and so on. But extrapolating to the population at large based primarily on the results of a single subject badly violates standard statistical practice. In the end, the Baker and Bellis data are statistically worthless and no scientific conclusions can be drawn from them.[24] The statistical methods and analysis used by Randy Thornhill, Steven Gangestad and Randall Comer, in their support of the first requirement of the female choice selection model, i.e., that women have orgasm preferentially with high quality males, are also suspect.[25]

There are also new data that bear directly on the feasability of another key mechanism at stake in the female choice hypothesis for female orgasm, the conditionality of orgasmic response. In a new study published by physiologist Kim Wallen and I, those women who reported orgasm with intercourse had significantly shorter distances between their urinary opening and their clitoris, (which appears in Figure 7.2 as the measure, 'clitoral-urinary meatus distance' (CUMD)), than did women who did not report orgasm with intercourse.

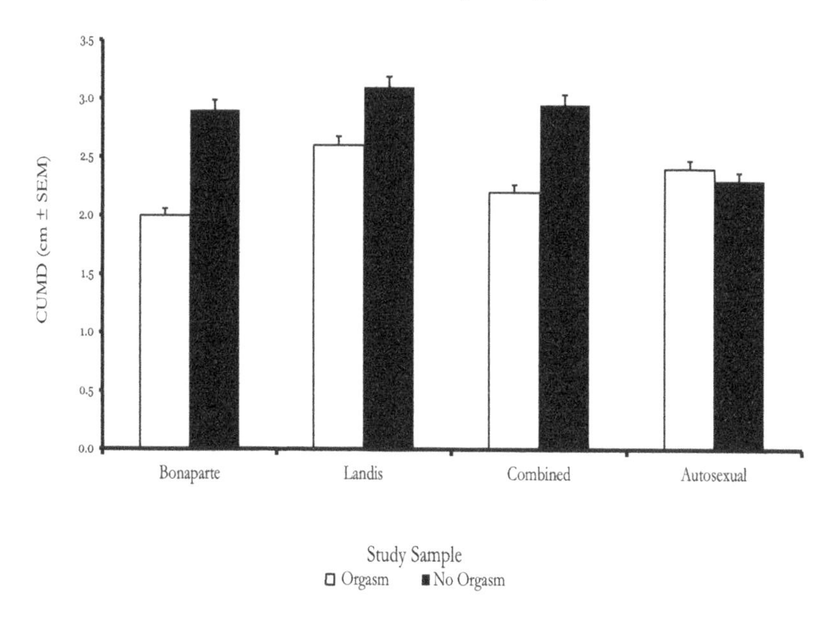

Figure 7.2: Clitoral-urinary meatus distance on y-axis, samples on x-axis, including women who have orgasm with intercourse (white bars), and women who do not typically have orgasm with intercourse (black bars), in the various study populations: Bonaparte, Landis, combined (Bonaparte plus Landis), and Autosexual (masturbation). Source: K. Wallen and E. A. Lloyd, 'Sexual Arousal in Women: Genital Anatomy and Orgasm in Intercourse', *Hormones and Behavior*, 59:5 (2011), pp. 780–92.

This difference in positioning of the clitoris and its relation to orgasm with intercourse was highly statistically significant. (The difference was greater than two standard deviations, and the r-value for the combined sample was 0.6.) There was no significant difference between the two groups with masturbation. We also found that this anatomical distance was strongly predictive of whether a woman had orgasm with intercourse.[26]

In other words, Wallen and I found that for the vast majority of women, anatomy strongly indicates destiny. Clearly, if her anatomy so strongly influences whether or not she has an orgasm with intercourse with a male, that leaves little room for the genetic quality of the male to influence the outcome of such intercourse. Obviously, if all selection needed was a bit of wiggle room to act, there is plenty of that available in this scenario; but that is not true in this particular case of sexual selection. According to the theorists articulating this type of female choice sexual selection model, the selection pressure of this type of selection scenario needs to be quite strong in order to produce any result in terms of evolution.[27] Thus, the anatomical relation we discovered makes such a scenario nearly completely unfeasible.

Developmental Account

Just because the adaptive accounts of female orgasm have largely failed, that does not mean that an evolutionary account is not available. An alternative, non-adaptive explanation for the evolution of female orgasm was first proposed by Donald Symons in 1979. Consider the problem of why men have nipples. Nipples clearly provide a reproductive advantage to *female* mammals, in that they ensure reproductive success by helping to feed the offspring. But there is no known contribution to fitness for the males.

The evolutionary explanation for the presence of male nipples is based on the development of the embryo, and the fact that nipples are adaptations in females. Males and females share the same embryological form at the beginnings of life – they start off with the same basic body plan, and only if the male embryos receive a jolt of hormones during the eighth week of pregnancy do any sexually distinguishing characteristics appear. The basic body plan includes nipples for both sexes because they are laid down early in development and the female needs them as an adaptation for nursing her offspring. Similarly, in males, orgasms are adaptations – they were actively selected for – but the females get them for free.

This sort of explanation is called 'developmental' or 'non-adaptive' – female orgasm is seen as a by-product of selection on male orgasm. The early embryo has a bi-potential genital tubercle, which turns into the head of the penis in the male and clitoris in the female. These organs with common embryological origins are called homologues. The tissues involved in orgasm for males and females are also homologues, including nerve tissues, erectile tissues and muscle fibres. So females get the functioning orgasmic tissues through this embryological connection and are often capable of having orgasms under the right conditions of rhythmic stimulation.

There is a variety of evidence supporting this by-product account of female orgasm – although it has encountered a great deal of resistance. Part of this is surely the name. Feminist sex guru Susie Bright commented that calling it the 'byproduct account' made female orgasm sound 'like a can of spam'.[28] So I am thinking of renaming it the 'fantastic bonus' account, which is much more accurate, after all.

Symon's selectionist/by-product account (selectionist for the males, by-product or 'fantastic bonus' for the females) accords well with the data about human female sexuality. Strikingly, females masturbate through direct stimulation of the clitoris – the homologous organ to the penis – and not through simulating intercourse, just as we would expect on the by-product account.[29] Similar stimulation to homologous organs yields orgasm for both sexes.

It also explains the low rate of reliable female orgasms with intercourse, which becomes quite predictable under Symons's developmental hypothesis; the reason these orgasms are so infrequent is the straightforward lack of the appropriate necessary rhythmic stimulation to the homologous organ to the penis, the clitoris. This needed stimulation can be provided through manual or mechanical rhythmic pressure on the clitoris during intercourse, which is how approximately half of highly orgasmic women achieve orgasm reliably with intercourse.[30]

Symons's general thesis is also supported by the nonhuman primate evidence. For instance, female stumptail macaques have been shown to have orgasms by hardwiring them up to detect the orgasmic response, showing that they have the distinctive contractions and other bodily markers characteristic of orgasm.[31] But many of the stumptail female orgasms occur during homosexual sex, as well as outside the estrus period, thus demonstrating no connection to fertility or reproductive sex, making these orgasms difficult to characterize as adaptive in any way.[32] Note that the by-product account does not deny that the *clitoris*, as an organ of sensation, has a crucial evolutionary function. On the by-product view, the clitoris almost certainly has been selected because it aids the female in sexual excitement and induces and prepares her to seek out and have intercourse. However, this reasoning does not extend to the use of these same tissues for female orgasm.[33] Orgasm is a special reflex that sometimes results from clitoral and genital excitement. General sexual excitement and the special reflex of orgasm are easily distinguished by most women, and just as easily distinguished by natural selection.

Finally, the disconnect between female orgasm itself and traits that might indicate that orgasm is an adaptation was demonstrated powerfully in a recently published study by Brendan Zietsch, Geoffrey Miller, Michael Bailey and Nicholas Martin. They examined correlations of such potentially adaptive traits and orgasmic activity in a population of nearly 3,000 women, finding zero to very weak correlations across all nineteen traits they examined, including libido, social class, orientation towards uncommitted sex, restrictive attitudes towards sex, marital status, age of first intercourse and lifetime number of sex partners.[34] None of the correlations had significant genetic components, thus undercutting any ascription of a fitness benefit to orgasm. There has never been any evidence linking orgasm to fitness, and this new study echoes this very significant absence.

Methodological Adaptationism

Let us turn to a large bias operating in this case. Earlier, I outlined how unconscious androcentric biases contributed to a mistaken assumption about the frequency of orgasm with intercourse, and a mis-evaluation of the evidence. Now, I would like to discuss how the theoretical bias of adaptationism has distorted both the theorizing and evaluation of evidence in this case. Many researchers who assume a leading role for adaptations in biology start with what philosophers of science call the heuristic or methodological adaptationist research question: 'what is the evolutionary function of female orgasm?' Other evolutionists, though, may ask: 'what evolutionary factors account for the form and distribution of female orgasm?' The possible answers to these evolutionary questions, which I will call the 'adaptationist' and 'evolutionary factors' questions, respectively, are relevantly and significantly different.

A basic theoretical background assumption emphasized when doing research on the evolutionary factors approach is that not every biological character is adaptive – that there exist alternative evolutionary explanations sometimes appropriate, such as phyletic explanations, by-product accounts, and accounts that cite correlations of growth. However, there is an important difference between paying lip service to this view and using it in actual research, as Stephen Jay Gould and Richard Lewontin famously argued in their 1979 'Spandrels' paper.[35]

Philosophers of science have always thought that the methodological adaptationist approach, one in which the evolutionist simply starts research into a trait by assuming that the trait has some evolutionary function, was fairly harmless, because it was assumed that if evidence arose that the trait was not adaptive, or might perhaps be a by-product, or constraint, then the biologist would take on that alternative account. As Peter Godfrey Smith wrote, 'when the hypothesis of optimality is investigated first, deviation from the optimum provides evidence that other factors are at work, and perhaps the nature of the deviation will give clues about where to look next'.[36] The issue here is exactly when the search for an adaptive explanation for a trait should be abandoned; but it is unclear when, in practice, this approach ever allows non-adaptive explanations to win the day.

Researchers studying female orgasm have been unable to make this switch from the adaptationist question to the evolutionary factors question, even when the evidence strongly suggests such a move. They even fail to see the by-product account of female orgasm as a distinctive alternative positive causal hypothesis, and as one that can have evidence in its favour. Therefore, they additionally fail to directly compare the by-product hypothesis against an adaptive hypothesis with regard to the evidence. Nearly all of these errors appear to result from the biases induced by methodological adaptationism, that is, the practice of assuming that the trait is an adaptation and asking what its function is. The view has been that this is a beneficial and harmless way to go about evolutionary investigation, but in this case, at least, it appears not to be so.

For example, for some adaptationists, the non-selective account is treated as the failure to find an evolutionary explanation, which they view as akin to scientific surrender. It is not seen as a positive evolutionary explanation at all because it is not an answer to our adaptive question, 'what is the function of this trait?' It is nonresponsive, because it says that the trait has no function. Leading animal behaviourist John Alcock, author of the most popular textbook on animal behaviour, takes this approach in his reaction to the by-product explanation of female orgasm. On his analysis, which is shared by Paul Sherman, a leading theoretician in the field of the evolution of animal behaviour and human evolution, the by-product hypothesis offers only a 'proximate' explanation of how women come to have orgasms. In other words, it explains how female babies grow up to have orgasms as adult women.

Alcock writes that

> proximate explanations of a biological characteristic do not make it impossible to ask whether the trait of interest contributed to individual reproductive success in the past or does so currently. If we were to discover the female orgasm occurred with positive effects on female reproductive success, we would gain an *evolutionary* dimension to our understanding of this trait that is not covered by *any* proximate explanation.[37]

Thus, the by-product account is not seen as an evolutionary account at all – it is not an answer to *any* evolutionary question about female orgasm, with supporting evidence and theoretical standing in evolutionary theory.

Similarly, David Barash, the author of the most widely selling textbook on sociobiology for a couple of decades, and his wife, Judith Lipton, write, regarding the impetus behind those favouring the by-product theory, that it involves 'a scientifically legitimate desire to explore all possible explanations for any biological enigma of this sort, including the "null hypothesis" that it might not be a direct product of *evolution* after all'.[38] Here, note the conflation of evolution with selection in this statement; again, the by-product explanation is not considered evolutionary.

However, the by-product theory is a positive causal evolutionary hypothesis. It proposes an alternate causal evolutionary theory for the form and distribution of the trait of female orgasm in the population, and thus is not a null hypothesis at all. We can see here some damage being done by methodological adaptationism, under which it is kosher to ask only the question: 'what is the function of this trait?' Because when an answer comes back apparently suggesting 'this trait has no function', this is seen as a non-answer, a scientific failure, which must be rejected as nonresponsive. Thus, even though methodological adaptationists such as Alcock, Sherman and Barash, present their adherence to their research programme and its attendant questions as perfectly harmless, and even as highly profitable and reasonable, we can see here where that programme goes astray. The difference between the questions, 'what is the function of this trait?' and 'what evolutionary factors account for the form and distribution of this trait?' turns out to be crucially important. A by-product explanation *cannot* be an answer to the first question, while it is perfectly acceptable for the second.

The fact that adaptationists see the by-product view only as a null hypothesis leads directly to their mistaken characterizations and inferences involving the view. For example, because they do not see it as an alternative positive causal hypothesis, they fail to see that it can account for the continuing existence of the trait over evolutionary time. Several adaptationists repeatedly complain that under the by-product hypothesis, female orgasm would fade away and deteriorate over evolutionary time, and would tend to disappear from the population. This notion has been advanced not only by Alcock (1998), Paul Sherman (1989), and David Barash and Judith Lipton (2009), but also by leading primatologist and human evolutionist Sarah Blaffer Hrdy (2005), and it is based on a misunderstanding of how the selectionist/developmental mechanism involved in the by-product account actually works.[39] Under that account, the basic muscle, nerve and tissue pathways involved in orgasm are shared by both sexes, and would be maintained in the female over the generations in virtue of the fact that they are strongly and continually selected in the male.

This case reveals that there is something peculiar and implausible about the methodological adaptationism advocated by so many biologists researching human evolution, namely, that the researcher is envisioned as, at some time in the middle of their research programme, abandoning their research commitments and explanatory practices, in the face of some facts or other, and then going along some completely different explanatory pathway. This is a lot to ask a researcher to do, and few – in fact, none – of the researchers involved with the evolution of female orgasm have been able to do that.

Hence, what is seen as and presented by biologists and philosophers alike as a benign bias embodied in methodological adaptationism turns out, I think, not to be so benign. Because such adaptationism begins by making salient only one type of evolutionary answer to the question of female orgasm, an adaptationist one, scientists working within the approach seem unable to switch questions and to engage in an unbiased and clearheaded way with the indirect selective and developmental alternative theory of female orgasm. What would it take for them to take no for an answer?

I am not saying that evolutionists should not search for adaptations or functions of female orgasm. I am, rather, highlighting some serious apparent risks of a particular approach to research into evolutionary causes. The presence of researchers like Symons who engage in their research using the more inclusive 'evolutionary factors' approach exemplify an available alternative method.

Feminist Critiques

Unfortunately, the fantastic bonus account of the evolution of female orgasm has been attacked by a number of feminists as being too androcentric, although some leading feminists, such as Susie Bright, as well as younger-generation feminists such as Tracy Clark-Flory from Salon, have seen it as a potentially liberating hypothesis. The immediate reaction of many feminists upon hearing the hypothesis – that a very important female trait is being 'reduced to a mere by-product' – seems to arise from translating the term into its everyday meaning rather than sticking to its scientific one, and also adopting a view that only adaptive traits are important. However, there is no reason to think that only directly selected traits are 'important'. Many traits that are not directly selected are considered extremely culturally important, such as refined musical ability, or the ability to write, read or use a computer. Similarly, even strongly selected traits, such as the swallowing reflex, are often devoid of cultural importance.

Much of the feminist reaction against the fantastic bonus account is based on substituting an everyday meaning of 'by-product' for the very different scientific one, and falsely equating what is important with what is naturally selected. We, as a culture, decide what is important, and its evolutionary historical genesis does not dictate what cultural attitude we should have towards female orgasm. Another way to express this point is to note that being a target of selection does not make a trait 'active' in any meaningful sense, nor does *not* being a direct target of selection – for example, being a by-product – make a trait 'passive'.

I am not a believer in deriving social norms from biological findings, but it is simply a fact that people routinely do just that. In this light, the by-product or fantastic bonus account of orgasm has a distinct advantage over other early accounts of female orgasm, because it expects no particular 'adaptive' set of responses to intercourse, and thus privileges none.[40] In other words, it casts all women as equally 'normal' in their orgasmic responses to heterosexual intercourse. The long struggle that feminists have undergone to separate the definition of women and their sexualities from their reproductive role can be partially achieved through this analysis.

Yet many feminists were repulsed by the fantastic bonus account, as Symons presented it. For example, Anne Fausto-Sterling, Patricia Gowaty and Marlene Zuk contrasted Symons's by-product theory unfavourably with David Buss et al.'s views about female orgasm in 1997:

> *In contrast, David M. Buss*[41] *notes that orgasm increases sperm retention; evolution, then, ought to select for men who stimulated female orgasm during or just following intercourse. In Buss's version of evolutionary psychology, women have much more agency than they do in Symons's.*[42]

One problem here is that Buss, in his textbook, *The Evolution of Desire* (1994), uses the scientifically hopeless Baker and Bellis studies as the basis of his explanation. I should note that both Zuk and Fausto-Sterling, despite my criticism of them, gave a very generous welcome to my book and also its harsh critiques of Buss's view, in their reviews.[43]

But in her review of *The Case of the Female Orgasm*, philosopher of science and feminist Lisa Gannett disagrees with the idea that the fantastic bonus account may allow a progressive view of female sexuality. Rather, she writes,

> if female orgasm arose as a developmental byproduct of male orgasm, and male orgasm arose as an adaptation because the pleasure experienced encouraged the frequent pursuit of sex, then we are presented with an evolutionary account of why – in Fausto-Sterling et al.'s words – males might be expected to be 'sexually predatory' and females might be expected to be 'passive'.[44]

Gannett and others have a potentially important objection to the by-product account, namely that it perpetuates negative stereotypes. Because, by hypothesis, orgasm does not contribute to reproduction, females are claimed to maintain a 'subordinate' role in sexual interactions.

Now, cast back to the risks associated with methodological adaptationism. Specifically, think of the way that it can narrow the search for adequate answers to a research question, that is, looking only for an adaptive function of the trait under consideration, or else failing altogether.

Under Gannett's approach, she implicitly searches only for candidate functions for female orgasm, and, faced with repeated failures, she sees the trait as negative, passive, 'subordinate' and lacking agency and 'autonomy'.[45] We can see by this that she is taking on the role of a methodological adaptationist, who values only explanations about functions. However, under the alternative evolutionary factors view, under which the fantastic bonus view falls, female orgasm is an actual causal trait on its own, and not negative, passive or one lacking agency. Gannett's deeply adaptationist assumption thus values a position that feminists have no good reasons to accept.

The by-product view may offer a very different picture of female sexuality. It is a mistake to conflate all of female sexuality into discussion of the orgasm itself. Females may be very interested sexually in virtue of their clitorises and the subsequent sexual excitability and interest in heterosexual intercourse as well as possible sexual activity with other females. As I mentioned earlier, the sexual excitement arising from the clitoris and surrounding tissues and the actual orgasm are *separate* features of the female sexual response, and sexual excitement occurs routinely in the absence of orgasm. The apparent lack of adaptiveness of female orgasm itself does not imply the lack of function of other aspects of female sexuality, especially the role of the clitoris in sexual excitement, which is considered an adaptation under the by-product approach.

Thus, what is known as 'proceptive' (rather than 'receptive') sexual behaviour – which includes interest in furthering sexual contacts and the pursuit of sexual behaviour – is suggested by the by-product view. Thus, also, the negative stereotype of the 'passive' female is not entailed by the by-product account of female orgasm, contrary to the fears of critics.[46]

Gannett claims that the by-product theory is inevitably tainted by association with other, possibly objectionable views of female sexuality, but gives us no good reasons for accepting such a claim.[47] The theory that copulation is a female service, and that orgasm is a by-product, are logically independent, as she snidely acknowledges,[48] but still claims that they are nonetheless 'mutually reinforcing propositions' in a web of evolutionary beliefs that Fausto-Sterling et al. are challenging. In the conclusion of this discussion, we get a comparison with Mill: 'Darwin is said to have reacted to John Stuart Mill's On the Subjection of Women by saying that Mill had something to learn from biology (Desmond and Moore, 572). Plus ça change'.[49]

So instead of reasons or evidence, we get innuendo in her attack on my position. Significantly, the by-product view and female service theory are *not* inevitably bound together and mutually reinforcing the way that Gannett and Fausto Sterling propose: in fact, nearly all of the scientists involved in this debate believe just the opposite, namely that some adaptive view – perhaps the female service view – of intercourse, and an *adaptive* view of female orgasm are bound together, so she's got the biology and the logic entirely wrong.

To review our results in this section: first, I have emphasized that there is no connection between the social importance of a trait and whether or not that trait is an evolutionary adaptation. In addition, whether or not something is a direct target of selection does not imply anything about whether it should be seen as 'active' or 'passive' in a meaningful sense for society. Importantly, on the by-product account, all women's orgasms are seen as equally expected and equally 'normal'; none are couched in terms of their superior functionality, as in the bad old days of Masters and Johnson, when women who did not have orgasm with heterosexual intercourse were called 'dysfunctional'. Also, on my analysis, the accusation that the by-product account of female orgasm somehow promotes a 'passive' or 'subordinate' picture of female sexuality seems to be founded on an adaptationist picture; in other words, it seems to be a view in which traits are 'autonomous' or can convey 'agency' only when they contribute to reproductive success. In addition, the view that only traits that are 'targets of selection' can be 'active' is also tied in with this adaptationist bias.

Conclusions

In sum, nearly all applications of evolutionary biology to the issue of female orgasm have been problematic – half of the theories offered fail due to androcentric assumptions, while the other half are faulty due to methodological adaptationism or other reasons, primarily sloppy and misleading methodology and conflict with the sexology findings. Only one – the by-product or fantastic bonus explanation – is currently consistent with and supported by the available body of evidence. We have been using sexology research as the fundamental evidence against which the evolutionary theories are tested, even though there may be serious questions regarding this evidence, because it is usually gathered by questionnaires based on past experience.

Here, it is essential to emphasize the conformity of results across more than 90 years of data-gathering concerning the frequency of orgasm with heterosexual intercourse, an example of which appears in Figure 7.1. These studies were done by researchers with very different theoretical biases and different backgrounds, with different expectations about the results. They used different methods, and had very different subject pools. Two of the large studies were done using statistically rigorous random sampling techniques from the general national populations of the United States and United Kingdom. Yet they all arrived at nearly identical results concerning the low rate of women having orgasm with intercourse on a reliable basis, as well as the 5–10 per cent of women who never have orgasm at all. These results are thus robust and consistent across testing instruments, data pools, research groups and measurement techniques, as well as being robust across eight decades of the twentieth century. Thus, we do have some window on many women's orgasms over the last century and the beginning of this one. These studies are capable of serving as the evidence against which evolutionary explanations should be compared.

Significantly, nearly[50] all of the latest empirical evidence which has appeared since the publication of *The Case of the Female Orgasm* on the subject is either consistent with[51] or strongly reinforces[52] the by-product theory of the evolution of female orgasm. The reinforcing evidence simultaneously directly and strongly undermines the most favoured adaptive accounts, despite evolutionists' clear bias in favour of such explanations.[53] This case has served as a clear and unfortunate example of long-standing male-centred bias affecting science, as well as an example of the effects of adaptationist bias distorting the examination and evaluation of the by-product theory and the evidence related to it. Adopting the fantastic bonus explanation of female orgasm might have profound social and medical consequences. If female orgasm is seen as having no particular evolutionary function, but rather as an evolutionary freebie, then many diagnoses of 'Female Orgasmic Disorder' would be out the window, as Zietsch et al. recognize, and women anywhere on the spectrum of orgasmic performance might be seen as normal. The issue of the evolution of female orgasm is still unsettled, and more research and evidence is needed to decide the case. But equally, we need more careful and thoughtful examination of the evidence that we already have, with all parties approaching it with open minds.

8 ALTRUISM AND SEXUAL SELECTION

Lucrecia Burges, Camilo J. Cela-Conde and Marcos Nadal

Introduction

The central idea of evolutionary ethics is that human morality – including both moral sense and moral codes – can be explained in similar adaptive terms. Human beings have the morality they do because, through the process of evolutionary adaptation, such a morality enhances survival and reproduction of human groups.

Moral ability is a necessary consequence of our biological evolution. Human beings are ethical beings because of their biological nature, which makes them capable of judging their own behaviour as either good or evil, moral or immoral. This is a consequence of a characteristically human intellectual ability that includes self-awareness, symbolic language and abstract thought. These intellectual abilities emerge as a result of an evolutionary process similar to those of closely related primate species, but they have definitively acquired a specifically human character.

In a well-known section of his *Descent of Man* (1871), Charles Darwin grounded the capacity to behave in a moral manner on a uniquely human feature: the moral sense.[1] He believed that this trait distinguished human beings from other living animals, though some, such as other primates, would exhibit it if their capacities could develop enough.

In the ethical naturalism inaugurated by Darwin, this notion of a moral sense linked morality to human nature. However, the specific mechanisms underlying the alleged moral sense were unclear. Darwin suggested that it emerged from the addition of sympathetic – i.e., emotive – impulses and the specifically human ability to reflect upon the consequences of our acts. Nevertheless, neither can the actual combination of these components be easily demonstrated, nor does the result seem compatible with the laws of natural selection. Even in Darwin's own writing, it was clear that altruistic behaviour (leading an individual to invest his or her own resources to maximize the adaptive fitness of another), as well other moral phenomena, was in opposition to natural selection. How could a mechanism that maximizes individual fitness favour altruistic strategies? Natural selection would probably eliminate any individual who tended to decrease his or her own fitness in favour of another's. In spite of this fact, the behaviour of certain organisms (among which we find amoebae, ants, rodents and humans) includes certain forms of altruistic acts.

The modern study of the biological foundations of social behaviour can be traced back to the flowering of a new approach to animal behaviour that grew from ethology during the 1960s and consolidated in the mid-1970s. This field came to be known as sociobiology. Edward Wilson described sociobiology as the systematic study of the biological basis of any social behaviour exhibited by animals and humans.[2] The most remarkable feature that sets sociobiology apart from ethology is the adoption of the gene's point of view and the development of a conceptual apparatus from this perspective. Prior to the consolidation of the actual field of sociobiology, many ethologists assumed that natural selection had favoured the appearance of certain behavioural patterns that were costly for the individual but favourable for the group. Hence, some aspects of animal social behaviour could be explained as individual sacrifices for a collective good. V. C. Wynne-Edwards made some of the most significant contributions to this way of thinking.[3] For instance, he argued that the altruistic renunciation of reproduction by number of individuals is responsible for the fact that groups of individuals or species that limit their growth as a function of environmental resources have higher survival rates than those that overexploit their habitats.

The criticism of group selection models elaborated by George Williams, as well as his convincing arguments in favour of more parsimonious explanations,[4] paved the way for the work of Edward Wilson and Richard Dawkins,[5] who championed the gene's point of view for the adequate comprehension of social behaviour. From this perspective, animal behaviour, especially social behaviour, was viewed as a means developed by genes to ensure their own transmission to offspring. As expressed by K. N. Laland and G. R. Brown, the body and its functions became mere vehicles for gene transportation and transmission.[6] However, this perspective did not make the explanation for altruist behaviour appreciably easier. If the objective of genes, and the social behaviours related to them, is to maximize their own possibilities of transmission to future generations, how can we explain the behaviour of individuals who reduce their own chances of survival and reproduction to increase others'?

In an attempt to explain these seemingly paradoxical acts, sociobiologists began using the concept of 'inclusive fitness',[7] and developed a non-individualistic model of evolution by natural selection: 'kin selection'. This notion was based on the fact that closely related individuals share copies of many genes. Hence, animals can increase the presence of those common genes in subsequent generations by favouring the reproduction of close relatives. Although altruistic acts of self-sacrifice to benefit another involve losing the opportunity to transmit one's own genes, they also increase the chances of transmitting other copies of those genes if the beneficiary is a close relative. In fact, Hamilton predicted that this behaviour would be selected if the cost to the altruistic individual is less than the benefit to the recipient or recipients multiplied by the probability that the recipient possesses the same gene. Thus, the closer the relative, the greater the sacrifice the organism is willing to make.

Robert Trivers introduced the notion of reciprocal altruism[8] as an answer to the question that arises immediately from Hamilton's arguments: how can we explain altruistic behaviour among non-related organisms? Trivers suggested that altruistic behaviour – which would initially be costly for the actor but beneficial for the recipient – could appear between non-related individuals that interact for extended periods of time. They would be especially likely to appear if there were a high probability of the altruistic act being returned by the other individual on a future occasion. Under these circumstances, over time, both individuals will have benefited more from their altruistic interaction than if they had not collaborated. In this case, the difficulty is to overcome the tendency of individuals to behave non-reciprocally, that is to say, to cheat. Reciprocal altruism is frequently observed in human beings, who have developed special procedures, such as altruistic aggression,[9] devised precisely to avoid the appearance of cheaters and to deter them from reiterating their behaviour.

Human Altruism

The term 'altruism' was originated by the French sociologist Auguste Compte, who wrote much about the development of altruism and the 'sympathetic instincts'. Altruism may be defined as unselfish regard for or devotion to the welfare of others.

In evolutionary theory, altruism means self-sacrifice performed for the benefit of others. An altruistic act is one that has the effect of increasing the chance of survival –'reproductive success' – of another organism at the expense of the altruist's. Ethologists are interested in altruism because it is a paradox for Darwinian theory. The paradox is this: by definition, altruists would be expected to have a lower reproductive success than that of selfish competitors. Altruistic behaviour should, therefore, disappear from the population. It should not exist, yet apparently it does.

Some people think that the answer to this paradox is that although altruism is disadvantageous to the altruist it makes up for this due to its advantage for the group or species to which the altruist belongs. Groups possessing altruistic members are less likely to become extinct than wholly selfish groups. Hence altruistic behaviour survives in the world.[10]

The implications of group selection, kin selection, and reciprocal altruism have been of enormous interest for the comprehension of animal altruistic behaviour. Granting that these models can successfully explain the altruistic behaviour of ants and rats, are they also useful to explain human altruism? To put it in other words: are we referring to the same phenomenon when we speak of altruism in both ants and human beings? The relation between moral altruism (exhibited by humans) and biological altruism (exhibited by other animals) is too complex to answer such questions with a simple yes or no. Several authors have underscored the numerous difficulties that arise when models and theories developed for the interpretation of the behaviour of hymenoptera are transferred to the study of human behaviour.[11] However, even accepting that such difficulties exist, we cannot agree with those who believe that there is no relation between biological and moral altruism.[12] Human behaviour is also the result of natural selection, so it falls under the definition of biological altruism that we noted above. By means of moral behaviour, humans reduce their own resources in favour of other individuals. Hence, 'moral altruism' is a special kind of 'biological altruism'. The same could be said of 'social altruism', referring to altruistic behaviours exhibited by insects belonging to the order hymenoptera. Natural selection has fixed extreme altruistic behaviour in at least four instances: hymenoptera (ants, wasps, bees, termites); parasitic prawns of coral sea anemones (*Synalpheus regalis*);[13] naked mole-rats (*Heterocephalus glaber*);[14] and primates (with humans as the best example). Thus, the true question is whether explanations for altruism in one of these special cases can be extrapolated to the altruistic behaviour exhibited by the others.

Moreover, since altruistic behaviour has appeared four separate times during the phylogenesis of multicellular animals. It would seem unreasonable to believe that extreme altruistic behaviour in those four sets of species was inherited from their most recent common ancestor. The trait is thus a homoplasy, meaning that its similarity among species is due only to convergence during separate adaptation processes. It has no significance regarding evolutionary relatedness. The evidence we obtain of the behaviour of any of those eusocial lineages cannot, therefore, be straightforwardly extrapolated to that of any other. The scientific success of the explanation of the evolution of the social behaviour of bees and ants does not afford many conclusions about humans. A different issue is that kin selection and other approaches constitute elegant mathematical models, capable of explaining the way in which a gene that promotes altruistic behaviours could be inherited. The speculations as to whether an allele capable of promoting cooperative behaviour is 'altruistic' or not clearly reveal the risks of confusing a purely biological concept with a common-sense one. An 'altruistic gene' or a 'selfish gene', as conceived by Dawkins,[15] should not be understood as possessing the same features that altruistic people do. But it is difficult to avoid the semantic load of language.

There is no doubt whatsoever that humans exhibit both selfish and altruistic behaviours. However, is our altruistic behaviour accurately described as the biological altruism conceived by kin selection and other models? It seems that parents actually sacrifice much for their children, for instance. Yet cultural deviations from this simple behaviour can introduce significant complexity. We mean to say that reducing human altruism to a simple behaviour, under the control of a few genes, is of little help to understanding the evolution of human moral cognition and behaviour.

Elliot Sober and Robert S. Wilson convincingly showed that the universe of human ethics is explained better by the model of 'group selection' than by models of individual selection.[16] This notion was already introduced by Darwin himself when, unable to explain the ultra-social behaviour of hymenoptera, he spoke of the adaptive advantages that a group of co-operators would have over a group of selfish individuals. However, this common-sense idea again seems to violate the principles underlying the mechanism of natural selection.

Altruism and Natural Selection

In the original Darwinian conception, natural selection refers to the individual adaptation of each organism. Let us grant for a moment that a group of altruists could actually adapt collectively, outcompeting other groups because its members engage in behaviours such as helping the sick and protecting each other against predators. Applying schemes from mathematical game theory, John Maynard Smith demonstrated that the adaptive strategy of such a group would not be evolutionarily stable.[17] If because of genetic mutations, recombination or immigration a selfish individual were to appear in this group, that individual would have a great selective advantage over the rest. Assuming simply that altruistic and selfish behaviours are determined by a single allele, the 'selfish genes' would eventually spread within the group, ending its cooperative nature. Maynard-Smith notes that a group of altruists can avoid the inconveniences caused by the presence of a non-cooperative individual if they are endowed with mechanisms capable of detecting and isolating any selfish individual that appears. However, this requires the members of the group to possess sophisticated cognitive mechanisms. Sober and Wilson showed that, unless the scope and content of such cognitive mechanisms are known, the explanatory power of group selection models cannot be properly assessed.[18]

Thus, one century and a half after Darwin's proposal of the mechanism of natural selection, we are faced with the same vagueness regarding the constituent processes involved in the human moral sense that Darwin himself expressed.[19] How can the combination of emotive and rational mental mechanisms that produce altruistic behaviour be understood?

The moral act is still described by authors like M. D. Hauser as a combination of rational and emotive components, which are hypothetically linked in the way that competence and performance are in a Chomskyan understanding of human language.[20] Hence, it could be possible to conceive something like a cerebral 'moral grammar' that guides 'our intuitive judgments of right and wrong'. This is just a conjecture at present, and does not explain how this grammar could have evolved by natural selection. Perhaps it actually did not, which would mean that it is not necessary to search for an explanation of how a trait such as altruist behaviour, which cannot be the result of the maximization of individuals' fitness, was selected.

All species are probably at least slightly 'behind' their environment, but this must apply *a fortiori* to *Homo sapiens*, a species whose ecological and social environment changes visibly and progressively year by year. This is one reason why we must be cautious when we are explaining modern human behaviour in terms of Darwinian evolution. Evolutionary psychologists have suggested that humans come into the world designed, in brain and body, for survival in the Pleistocene plains of Africa. What that brain does today might bear some relation to the roles for which it was originally selected, but the connection is fragile and it needs to be interpreted with subtlety and care. The human brain has taken off on a non-genetic evolutionary trip.

Some authors have suggested that human traits like language and moral sense could perfectly be regarded as by-products of unrelated adaptive episodes that caused certain neural modifications.[21] As we have already mentioned, Darwin stated that any animal with enough social instincts 'would inevitably acquire a moral sense or conscience, as soon as its intellectual powers had become as well developed, or nearly as well developed, as in man'.[22]

Michael Ruse has argued that morality is a direct product of biology, something that came about directly because of natural selection.[23] In Ruse's words morality is an adaptation.[24] F. J. Ayala believes that in the Darwinian conception, the moral sense is a necessary consequence of high intellectual powers. He thinks that, if our intelligence is an outcome of natural selection, so is our moral sense. Ayala believes that Darwin's statement further implies that the moral sense is not by itself directly conscripted by natural selection, but only indirectly as a consequence of high intellectual powers. In Ayala's view, morality is an exaptation.[25]

Is altruism an adaptation or, rather, and exaptation? Evolutionary biologists define exaptations as features of organisms that evolved because they served some function, but which were later co-opted to serve a different function, which was not originally the target of natural selection. The new function may replace the older function or coexist together with it. Morality is an adaptation in the sense that it contributes to the biological success of our species, but it is an exaptation, rather than an adaptation, in the sense that it was not directly promoted by natural selection.

If the moral sense is not a direct result of natural selection, what mechanism could explain its appearance as a consequence of high intellectual powers, or of any other cause, for that matter? Taking these hypothetical thoughts further, could the moral sense be a result of sexual selection?

Altruism as a Result of Sexual Selection

Darwin considered it difficult to explain the evolution of embellished, ornamental, colourful and flashy traits that appear to lack a clear utility, such as the male peacock's tail, by means of natural selection. His observation that many of these traits were related to mating processes led him to elaborate his theory of sexual selection, which he defined in *The Descent of Man* as the success of certain individuals over others of the same sex, in relation to the propagation of the species.[26] Darwin believed that such an advantage was the result of two possible types of competition: one is between individuals of the same sex, generally the males, in order to drive away or kill their rivals, the females remaining passive; while in the other, the struggle is likewise between the individuals of the same sex, in order to excite or charm those of the opposite sex, generally the females, which no longer remain passive, but select the more agreeable partners.[27] It must be noted that sexual selection cannot be considered as distinct from natural selection. Ethology and sociobiology provided evidence that such conspicuous traits as the male peacock's tail have an adaptive significance. However, sexual selection is regarded as a special kind of mechanism involved in certain selective episodes that drive newly evolved traits.

Darwin believed that sexual selection had profoundly marked the evolution of humans, and he stated that this process not only shaped physical traits, but also cognitive processes and behaviour patterns:

> He who admits the principle of sexual selection will be led to the remarkable conclusion that the nervous system not only regulates most of the existing functions of the body, but has indirectly influenced the progressive development of various bodily structures and of certain mental qualities. Courage, pugnacity, perseverance, strength and size of body, weapons of all kinds, musical organs, both vocal and instrumental, bright colours and ornamental appendages, have all been indirectly gained by the one sex or the other, through the exertion of choice, the influence of love and jealousy, and the appreciation of the beautiful in sound, colour or form; and these powers of the mind manifestly depend on the development of the brain.[28]

Darwin believed that sexual selection played a major role in the evolution of humans and the divergence among distinct human populations. Mate preference is seen as being subject to genetic influence while much modelling of sexual selection assumes genetic influence acting on the preferred trait.

In the present work we examine the merits and limitations of the proposal that altruistic traits evolved as a result of sexual selection under the special conditions of human evolution. A. Zahavi was the first one to suggest that altruistic traits might be linked to sexual selection. He proposed that altruism originated as a 'handicap' that evolved because it gave a 'costly' and therefore accurate 'signal' of the phenotypic and genetic quality of the altruist to others. The 'direct phenotypic benefit' mechanism envisages the sexually selected trait itself as offering an advantage to the female and her offspring.[29]

During Darwin's time, the theory of sexual selection did not receive the same amount of attention as the notion of natural selection. Nevertheless, as A. Paul noted, research on the mechanisms underlying sexual selection and mate choice thrived during the last quarter of the twentieth century, leading to interesting theoretical and empirical developments. The main shift was related to the role of males and females during mate choice: the notion of aggressive males competing to access passive females was gradually abandoned in favour of a view in which both sexes actively prefer and choose potential mates.[30]

As we have seen, several different models have been put forward to explain altruistic and other moral behaviour in humans and other social species. These models are not necessarily incompatible. In fact, claiming that one can explain human moral behaviours by reference to a single evolutionary mechanism would constitute a great over-simplification. In this spirit of arriving at a comprehensive and integrative view of the evolution of altruism, fairness and other moral behaviour, some authors, such as I. Tessman and G. F. Miller, have contended that there is evidence that some human moral behaviour evolved through sexual selection to serve display functions, just as sexually attractive physical traits and ornaments serve as signals of health, fertility and longevity. These authors have not stated that sexual selection created moral behaviour. Rather, they claim that sexual selection transformed a set of primate cognitive mechanisms into our uniquely human moral virtues. Also, they have not claimed that sexual selection was the only force involved in this transformation. Its effects must have intertwined with those caused by the other mechanisms mentioned above.[31]

Miller has presented his arguments in the greatest detail. He begins by reviewing evidence suggesting that many moral traits are sexually attractive. While some moral virtues may be attractive in themselves, the attractiveness of others resides in their function as signals for other desired traits. They may serve as indicators that their possessors are capable of having lasting cooperative relations and of investing in offspring.

From this point of view, moral virtues and behaviours constitute costly signals. Costly signals are indicators, resulting from natural selection, that usually advertise good genes, good parenting abilities, good long-term relation abilities, or a combination of these. Moral behaviour is a trait exhibited prominently during courtship. During this process, the potential partner evaluates such moral behaviour as generosity, sincerity, empathy and self-control. Evidence for moral flaws, such as greed, envy, cheating and lying, is also carefully sought.[32]

It is argued that certain personality traits, mental health traits and intelligence have a moral or almost moral rank. For instance, given that conscientiousness and agreeableness, two personality traits, are highly valued in long-term mates and are highly associated with partner and parenting traits, Miller concludes that they are most likely to have been shaped as moral virtues by sexual selection.[33] Also, it is posited that sexual selection acted upon the evolution of intelligence and mental health, given the fact that statistical tests have shown that they predict a broad range of righteous behaviour. Models of the evolution of morality based on sexual selection need not posit the presence of moral traits only in one sex, because both male and female humans actively choose their partners.

Miller argues that sexually selected moral traits should exhibit most of a set of twelve features. Their simultaneous appearance in relation to a moral trait would be a clear suggestion of the role of sexual selection in moulding such a trait. There are three groups of such features: genetic, phenotypic and those related to the sexual selection of moral traits.[34]

Genetic features

(1) Moral virtues should be genetically heritable and expressed mostly in adults.

(2) Interbreeding and increased paternal age should have negative effects on moral behaviour due to harmful homozygous mutations and increased mutation load in sperm, respectively.

(3) It is expected that alleles that reduce moral behaviour appeared fairly recently, and have not yet been eliminated by the effects of sexual selection.

Phenotypic features

(4) Moral behaviour should be conspicuously displayed during courtship.

(5) Moral behaviour should represent a significant cost to the producer and they should correlate with other fitness indicators, such as mental and physical health, intelligence, body size and symmetry, and so on.

(6) The absence of moral behaviour should correlate highly with neurodevelopmental disorders.

(7) Males should exhibit greater variance in their moral behaviour than females, given the level of polygyny observed in our species.

(8) Conspicuous moral behaviour should peak during young adulthood, when the effort invested in mating is at its peak.

(9) It is expected that those individuals exhibiting low levels of sexually attractive moral behaviour should attempt alternative mating strategies, including short-term opportunistic mating, deception, harassment or coercion.

Features associated with sexual selection

(10) Individuals expressing moral traits should be highly regarded by potential mates, who are expected to actively test whether potential mates exhibit such traits.

(11) It is expected that individuals who frequently exhibit moral behaviour should mate with each other, and that individuals showing fewer moral behaviour traits have no other choice than to mate among themselves.

(12) Rivals of the same sex are expected to criticize each other based on moral flaws, such as lying or cheating, whereas gossiping about potential mates with friends should revolve around the mates' moral character.

Although Miller's approach is currently the most comprehensive effort to explore the influence of sexual selection on the evolution of moral behaviour, it still faces some important challenges. The challenges are to: (i) the way Miller and Tessman, among others, have conceived the mechanism of sexual selection; (ii) the breadth of moral virtues they consider; (iii) the relation between the phenotype and genotype of morality; and (iv) the unclear nature of the mechanisms that evolved along the human lineage.

(i) Although Miller and Tessman speak of the relation between moral behaviour and sexual selection, as Paul has so clearly shown in his review,[35] there are a number of diverse sexual selection mechanisms. First, there are various mechanisms related to mate competition, such as endurance rivalry (extending periods of reproductive activity), scramble competition (locating potential mates before possible rivals), contest competition (using display or combat to eliminate rivals), reproductive suppression (inhibition of rivals' reproductive functions through endocrine mechanisms), sperm competition, or alternative mating tactics. Second, sexual selection encompasses mate choice mechanisms, which can be classified in three groups: precopulatory choice (behaviour that reduces copulation with certain potential mates); postcopulatory choice (female selection of different males' sperm after copulation); and postfertilization choice (using selective abortion or investment in zygotes, embryos or young). Third, Paul notes male sexual coercion as a different mechanism of sexual selection.

Attempts to explain the evolution of moral behaviour by means of sexual selection have not been explicit as to which of the mechanisms, or combination of mechanisms, they actually refer. Nevertheless, from these authors' work, it can be deduced that they specifically regard mate choice as the preferred mechanism related to moral virtues. H. Kokko and colleagues defined mate choice as the outcome of the inherent propensity of an individual to mate more readily with certain phenotypes of the opposite sex (i.e., mating preference or bias) and the extent to which an individual engages in mate sampling before deciding to mate (i.e., choosiness).[36] In this sense, both male and female humans are expected to prefer mates who exhibit moral behaviour as part of their phenotype, and to spend time assessing such traits in potential mates. It remains to be shown whether sexual selection mechanisms other than mate choice are relevant to the evolution of moral behaviour and how these diverse mechanisms interact among each other.

In addition to there being several mechanisms of sexual selection, there are different models that attempt to explain mate choice.[37] The direct benefit model assumes that the choosy individual obtains resources such as food, protection or parental investment. However, some authors have posited that the true benefits are actually indirect, and that the choosy individual gains only good genes for their offspring. Finally, there are authors who have proposed non-adaptive models of sexual selection, arguing that traits could be favoured merely as by-products of other processes. Miller and Tessman seem to favour indirect benefits models. Among these, they prefer the handicap view that moral behaviour is costly and, hence must be honest signals of high heritable quality.[38]

Again, it remains to be explored whether moral behaviour susceptible to sexual selection could be explained by either direct benefits models or non-adaptive models. In fact, there is no general agreement as to the superiority of any of these conceptions of mate choice. Kokko noted that the importance of indirect benefits has been questioned in many instances. Given that a relatively long series of processes is required to achieve the genetic benefits, profit can be minimal if one of the events turns out wrong, making choosiness a strategy that could reduce benefits in the long run.[39] Paul noted that mate preference seems neither to be species-specific nor uniform across the individuals of the same species, given that there is a fair amount of variation among individuals' choosiness and preferences.[40] Additionally, it is not clear that what is best for one individual is also best for another one. An individual's choice could be based on the search for potential mates whose genes constitute good complements for his or her own, and on the avoidance of homozygosity. This would lead to preference for mates carrying different alleles to one's own.

(ii) The second general issue concerned the range of moral virtues considered by Miller when he argued for the quasi-moral status of certain personality traits, mental health traits and intelligence.[41] However, the conclusion that agreeableness and conscientiousness were moulded into moral virtues by sexual selection because they are highly regarded in potential long-term partners and associated with positive partner traits should be taken with caution, given the conceptual nature of these personality dimensions. It has been pointed out in several occasions that the big five dimensions (openness, extroversion, conscientiousness, agreeableness and neuroticism) portray personality at the highest level of abstraction. Each of them, including agreeableness and conscientiousness, encompasses a fairly large amount of specific personality traits.[42] Hence, the big five should probably not be seen as personality traits, but as terms representing sets of concrete personality traits. It thus seems that research into the relation between moral behaviour and personality would profit more from a lower, more specific, level of analysis.

Miller noted that potential mates value agreeableness and conscientiousness positively because they predict good partner and parenting behaviour. However, agreeableness is defined as a pro-social and communal orientation towards other individuals, the presence of altruistic behaviour, tenderness, trust and modesty. Conscientiousness, in turn, describes socially prescribed impulse control that facilitates task- and goal-directed behaviour, such as thinking before acting, delaying gratification, following norms and rules, and planning, organizing and prioritizing tasks.[43] Thus, rather than both personality dimensions being preferred in mates because they predict attractive behaviour, it seems that they actually summarize a broad variety of behaviour that is attractive to potential mates, rendering the level of personality dimension superfluous. Probably the same can be said about intelligence and mental health.

(iii) We turn now to the third challenge, the relation between the phenotype and genotype of morality. Miller's proposal rests on the assumption that there are certain 'genes underlying virtues'. However, recent research concerning the genetic bases of language raises serious questions as to whether we can ever expect to find such clear-cut association between moral behaviours and specific genes. The gene FOXP2, the mutation of which is known to cause a severe speech and language impairment,[44] seems to be involved in the development of corticostriatal and olivocerebellar circuits, including the caudate nucleus and putamen, crucial to sequencing orofacial motor activity. However, there are serious doubts as to the language specificity of these neural circuits. They actually seem to be involved in a number of motor tasks.

Additionally, FOXP2 is not only expressed in the brain. It also seems to play a crucial role in guiding the development of other tissues – including lungs, some organs in the digestive apparatus and the heart, as in various tissues of the adult organism.[45] Back to the evolution of morality, the expectation of identifying genes which are exclusively related to the organization of neural structures underlying moral behaviour is probably unreasonable. We must expect the genetic underpinnings of complex cognitive processes to be an intricate pattern of relations among various non-specific genes.

With regards to the heritability of the two personality dimensions discussed by Miller, agreeableness and conscientiousness, results from twin studies show that these traits are actually among those influenced most highly by a person's environment.[46] Furthermore, in line with the notion that the level of abstraction at which the big five personality dimensions are defined make them too general to relate to a genetic basis susceptible to sexual selection, K. L. Jang and colleagues concluded:

> Our results suggest that genetic and environmental effects are not uniform across all facets of a dimension. For example, their data suggest that not all facets of Conscientiousness are influenced to the same degree by genetic factors. Individual differences in Order, Self-Discipline, and Deliberation appear to be largely determined by environmental influences. The implication is that some of the broad dimensions may not be etiologically homogeneous.[47]

Hence, the pattern of heritability of behaviours related to morality also seems to be a complex issue that requires further detailed research.

(iv) The fourth challenge facing any model dealing with the evolution of moral behaviour is the fact that other primate species share with us certain behaviours that have been considered the building blocks of human morality. There is a wealth of data indicating that our close primate relatives are endowed with complex social cognition, engage in reciprocal exchanges, have a sense of equality, have conflict resolution mechanisms, are capable of consolation, show sympathy to others and express other traits that are related to moral behaviour.[48] Hence, not all forms of human moral behaviour appeared solely at some point during human evolution. Our morality did not appear in a vacuum; it is the result of modifications and additions to an initial state of moral behaviour, that which we share with other primates. Thus, approaches to the evolution of human moral behaviour, whether they are based on natural or sexual selection, should take this fact into account.

Conclusions

In the present work, we have shown why the evolution of altruism and other forms of moral behaviour is not satisfactorily accounted for by means of natural selection alone. Darwin's main evolutionary mechanism is better suited to explain the phylogenetic history of traits that maximize individual fitness than those that increase the fitness of others. Models and approaches such as kin selection, reciprocal altruism and group selection, which aim to provide the theoretical underpinnings of the evolution of moral behaviour, assume that such behaviour does in fact afford benefits, albeit indirect ones, to the individual, its descendants or the genes it carries. We have briefly sketched these alternative views, underlining their explanatory power and noting each of their weaknesses.

Sexual selection has recently emerged as a strong complementary explanation for certain moral phenomena and their evolution. Sexual selection operates at diverse biological levels, though, in relation to morality, the most relevant of these seems to be precopulatory mate choice, or the restriction of all possible mates to a small set exhibiting certain features or behaviour. Theorists have favoured the perspective that choosy individuals receive indirect benefits, in that the preferred traits are highly costly and, hence, honest indicators of great heritable quality. In essence, it is argued that humans' moral behaviours were shaped by both sexes' preference for mates that exhibited them because they signalled good genes, or good mate or parenting abilities. We believe that this constitutes an appealing and original approach, which has proven to have a great heuristic value, suggesting novel and interesting questions. However, systematic research on the role of sexual selection on the evolution of moral behaviour is fairly recent. This is why there are still many issues to be resolved and limitations to overcome. These have to do mainly with the definition of moral traits, the way they are related to human genetic constitution, the interaction between sexual selection and other mechanisms, and the way the models regard the initial state of the evolution of moral traits. We have suggested some research avenues that might be helpful in the clarification of such issues and that could stimulate further work in this area

9 THE ROLE OF SEX AND REPRODUCTION IN THE EVOLUTION OF MORALITY AND LAW

Julia Sandra Bernal

Introduction

As human beings consider themselves members of a biological species, we can offer explanations of our origin and nature based on the theory of evolution. Such explanations can include the role of sex and reproduction in the origin of morality and the law.

The theory of evolution starts from the fact that organisms change through time in a process of variation, adaptation and selection where success is measured by the biological efficiency of individuals to leave descendants that in turn will also leave descendants. Individuals are active agents in their environment; they react, behave and interact with other individuals of the same or different species in an adaptive way. This drive to live has been classified by biologists as *selfish* and it is manifested in two very different forms called *positive selfishness* and *negative selfishness.*

We talk about positive selfishness when, through interaction, organisms obtain mutual benefits; these kind of relationships are in general called symbiotic. But, when through interaction with another, an organism obtains benefit at the cost of harm to the other, we speak of negative selfishness. These types of relationships are generally called parasitic.

R. Wright, who applies the theory of games to the biological evolution of organisms, states that, when organisms interact, they either follow zero-sum dynamics, where one wins and the other one loses, or non-null dynamics, where two organisms or entities benefit from the mutual interaction, although not necessarily through cooperation. From this point of view, when two organisms can reciprocally improve their survival and reproductive perspectives, they are in a situation of non-zero sum, while when their interests are opposed, the dynamic is of zero-sum.[1]

The deductive hypothesis that explains the fact that only a fraction of the organisms survive long enough to leave descendants, and that those who survive have adaptive variations for a given environment, which are transmitted to their descendants as if there was a filter mechanism, is what Darwin called 'natural selection'.

Darwin took into account that the essential elements or factors in the biological evolutionary process were, besides the organism, the environmental conditions where it dwells: abiotic (the different climatic, chemical and physical environment variables) and biotic (the organisms which share the habitat) and the reciprocal loop causal relations between them; so that the differences in the capacity of organisms to face selective pressure from the environment and the variables influencing their survival and their successful reproduction are the ones which determine, in the long term, their biological efficiency.

The above mentioned leads to the conclusion that organisms existing today, humans included, are the result of a biological evolutionary process from ancestors that survived long enough to have descendants. This biological evolutionary process is one of conservation and change at the same time; of constrictions (physical, chemical, physiological, anatomical and ecological), and also of alternatives and possibilities. Evolution is not necessarily towards complexity: bacteria, for example, have remained reasonably stable throughout evolutionary history. We cannot conclude either that in evolution there is a trend to progress, since evolutionary change is not an avenue to go from imperfect to perfect or from the lowest to the highest; what evolutionary change reflects is efficiency and the luck of organisms to live as autonomous individuals in their interactions with their environment.

Social Species

One of the ways individuals have survived is through cooperation while circumstances are adverse, such as in extreme weather conditions, or when predators pose a risk, or when localizing or obtaining food resources is difficult. Beyond simple positive interactions, it has been found that individuals of some species coordinate their behaviour and cooperate with each other in a reiterative way; and for this reason, they remain together for a long time. This has been the origin of sociability in several species.

Starting with the premise that individuals perform behaviours that contribute to their biological efficiency, behaviours defined as social can be objects of selection when the act of remaining together, the act of living in a group, helps them guarantee their biological efficiency, in spite of greater competition for food resources and mating and higher probabilities of getting diseases and parasites. Thus social individuals follow strategies that guarantee or increase their biological efficiency. Sociability can evolve when the advantages of being sociable are greater than the disadvantages for the individuals, i.e. when the cost–benefit relation has a positive balance for benefits, which is interpreted as biological efficiency. As E. O. Wilson says 'the most elaborated forms of social organizations are, at the end, vehicles of individual wellbeing, in spite of external appearances'.[2]

One of these efficient strategies has been related to reproduction in what has been called 'parental investment'. It has been documented that in some vertebrate species whose offspring are born immature in their development (altricial), and/or are highly vulnerable to predators, and/or have difficulties getting food by themselves, the parents cooperate in their raising. Because of such cooperation, the survival and reproductive possibilities of the offspring increase, which in turn increases the biological efficiency of parents.

It has also been found that individuals with kinship relations cooperate, helping defend the offspring, sharing food or alerting others to danger. These cooperative behaviours have been selected as they represent an adaptive advantage since they contribute to the reproductive success of those individuals' nearest relatives, thereby increasing reproductive success in an inclusive way by increasing the reproductive efficiency of the shared genotype of the cooperative, which has been denominated 'inclusive biological efficiency'.

In the case of social mammals, a key to their social evolution seems to be in breast feeding, since the offspring depend on their mothers during a substantial period of their first development in such a way that the group mother-descendant could be the nuclear unit from which different degrees of sociability develop, together with other factors such as population density, access to food resources, defence from predators, and so on.

All this leads to the claim that the selection of cooperative behaviour between non-related individuals has evolved at population level in species where individuals had patterns of parental and alo-parental investment behaviour and that, due to ecological biotic and abiotic pressure especially related to defence and nutrition aspects, these patterns spread among non-related individuals with whom the possibility to interact was frequent.

The behaviour of active sharing has been observed in *Cebus apella* (capuchins monkeys) at group level. This behaviour could have come from the group exploitation of a wide variety of plants and animals, which required the strength of the group and, in some cases, the use of tools.[3]

Sociability is a common trait in most monkeys and, therefore, it is very probably an ancestral trait that evolved because it offered adaptive advantages. In studies with primate gregarious species there was found a correlation between the benefits of sociability with the biological efficiency of individuals, taking into account factors such as habitat, the optimization of food resources and their distribution, the reduction of the possibility of becoming prey, the possibility to mate with a large number of partners or to increase the probability that the offspring survive until they reach sexual maturity and reproduce.

A strategy studied in chimpanzees is reciprocal altruism behaviour in four important services of social life: help in conflicts, delousing, sharing food and sex. These exchanges have self-regulating mechanisms, which include the ability to weigh the pros and cons in making a decision, for example, realizing that helping someone in a conflict means to opposing the other.

The emotions already present in mammals, which help them respond and communicate with other individuals in an adaptive way in social primates, play the role of self-regulation of individuals starting from the knowledge of their role within the group. Although the idea that they suffer from a sense of guilt is still in dispute, it has been established that, apart from their knowing and interiorizing social rules associated with their social role, they respond in ways that show they are aware of having violated a code. Subordinate *rhesus* monkeys that have mated with females in the absence of the Alpha male later show submissiveness and try to avoid him when in his presence.[4]

There also seems to exist in the group what De Waal calls a sense of social regulation: rules about proper behaviour, expectations about the treatment the individual or others should receive and guidelines about how resources should be shared. When reality fails to meet these expectations in detriment of the individual or others, there is a negative reaction: confusion, surprise, anger,[5] sorrow, revenge and 'moral sanction', which are part of the reciprocity system. It has been observed that when dominants mistreat others excessively, the others react as if an accepted limit had been trespassed, and this suggests, for De Waal, a sense of justice and impartiality.

Evolution of Our Social Species

The traits that have defined the *Homo* genus, more advanced bipedalism, more skilful hands, more developed and asymmetric brains, the use of carved stones and increase in meat consumption, seem to reflect a survival problem in the gathering of food resources and, therefore, a possible competition in an increasingly dried environment, which put the selection of characteristics and adaptive behaviour under challenge in the different species of hominids.

My hypothesis is that during the hominization process a directional social evolution process was generated at the population level that increased the biological efficiency of our ancestors.

Three essential factors led the evolutionary process of our species: *bipedalism*, *food* and the *helplessness of newborns.*

Bipedalism seems to indicate a selective advantage of omnivorous species,[6] which might have been the capacity to access a wide range of resources in a mosaic environment of forests and prairies that limited one another.[7] The combination of the abilities of walking on the ground and climbing on the trees allows the hands to be free while walking and the possibility of carrying tools and food, collecting fruits from low bushes and exploiting new food sources by capturing newborn or dead animals and taking them to more protected areas.[8] The ability to climb trees gave them an advantage over some predators: to spot them, evade them and to construct nests to sleep safely.

On the other hand, given the anatomical characteristics of our ancestors and the habitat they shared with other species,[9] it is feasible to suppose that the dynamics followed by the individuals in the population were of non-zero sum, i.e., that they developed joint activities with other members of the same population and that these activities were efficient to survive. The findings of skulls of different hominids from three to four million years ago and the studies of brain moulds seem to indicate that the greater capacity of communication between individuals and of elaboration and manipulation of instruments may be the result of a collective cooperation strategy in the gathering of food and defence against predators.

It is also probable that wandering through open spaces in the savannahs was not a solitary activity and that they used stones or sticks to defend themselves as savannah's chimpanzees do today; it is also probable that with a more varied diet, and in a savannah habitat with many predators, males' cooperation would prove more efficient than acting alone when looking for energetic food such as meat, and that they would share the product as has been reported in the studies of Tahi chimpanzees,[10] and that occasionally they would exchange meat with females for other energetic products such as seeds or for sexual intercourse.

On the other hand, females must have lived in more restrictive areas of forests where they were more familiarized with the distribution of food resources and the location of predators, so they could protect their offspring. It is also possible that, having to face competitors or predators, there was cooperation in the community 'nesting', i.e., that females would raise offspring together and cooperate: while some went to gather food, others would stay and take care of the offspring communally. Reproductive synchronization could have contributed to this cooperative behaviour, as it is seen in some bird species.

The model of our ancestors that is slowly emerging is that of short and skilful individuals with an omnivorous strategy for scavenging fresh, or occasional hunting, prey, for fishing, for collecting seeds and fruits, and for extracting roots and tubers; with the technical ability to use tools in obtaining resources of all types, and in general with a set of characteristics belonging to both fruit-eating and carnivore species.

This means that they had cognitive abilities to remember the best sources of fruit, the harvest season and the times of scarcity, and that they also had the strategic abilities of carnivore species such as courage, slyness, cooperation and coordination to avoid competitors of other species with greater anatomical strength, and the ability to improve defences against predators such as the sabre-toothed tigers.

The third crucial factor in the evolutionary process of our species is the so-called 'helplessness of newborns'. Although the bipedalism of our female ancestors did not originally seem to be a problem when giving birth, with time, a greater specialization in bipedalism and the increase in body size and the head of the offspring brought several inconveniences at the moment of childbirth.

The modification in pelvic structure derived from a greater specialization in biped locomotion represented a clear limitation in having a viable childbirth for offspring with larger brains. It seems that the evolutionary solution consisted in offspring born prematurely and with a faster brain postnatal growth combined with a slow somatic growth. In other words, surviving offspring likely had a variant of some regulation genes that altered the developmental rate;[11] a process of heterochrony that gave different body systems and organs a slower development after birth, which meant that the surviving, premature offspring were the most immature and helpless.

The main evidence for these modifications and their physiological consequences are found in the *H. erectus - H. ergaster species*, in which cranial capacity reached 800–950 cubic centimetres, which in turn rendered the passage through the narrow pelvic channel of the mother more difficult; although it is probable that the process had started before.

As we can see, this evolutionary solution meant a compromise: the fact that the newborn is in a neurological premature state and helpless required adjustments in the mother's behaviour. Thus emerged a longer period of child rearing. In addition, the increased metabolic consumption of a larger brain led to a larger energetic intake for its development, which in turn meant an increase in the nutrient quality, including a greater number of proteins and fat to allow for an energetic balance. Obtaining a fat and protein rich diet through the exchange of different food products would have reinforced cooperative behaviour and could have led to a change in social interactions, which in turn would have increased biological efficiency at the population level.

According to M. Domínguez, if we had followed the same behaviour of the other primates of having new offspring right after having weaned one, and with the female in charge of the offspring, the Homo genus would have never been able to survive, since the prolongation of the development that allowed brain enlargement and longer periods between childbirths, as weaning was longer, would have resulted in a lower reproductive rate.[12] For this reason, a strategy that increased the biological efficiency appeared: having more children and/or ensuring that those who were born had enough time to reproduce. Selective pressure centred then in getting the offspring to survive with the necessary help of other individuals.

One way of increasing biological efficiency could have been through the help of other female relatives of the mother, as for example the adolescent sisters, which as in chimpanzees and bonobos learn mothering by 'playing'. The so-called 'grandmother hypothesis' seems to explain the prolongation of the life expectancy of human females beyond menopause, for such women could be in charge of gathering tubers they would later share; or they could be in charge of the offspring while the mothers gathered food resources. In any case, they would increase their own fitness since the survival margin of descendants widens until they also reproduce and enables mothers to have more children.

From the initial male cooperation in defence of their territory, they moved on to collective scavenging and hunting and the sharing of the loot; and from an exchange of food resources with the females for sex or other reasons, they moved on to actively sharing resources, which could have been a vital factor in survival, since a better nutrition for the breastfeeding mothers led to a better nutrition for the offspring. Base campsites or settlements from 1.8 million years ago could have served to protect the young and as centres where individuals would get together temporarily to share or exchange food supplies.

Little by little they could have gone from an exchange of food for food or sex to an inter-gender division of labour and active sharing of resources leading to a real female–male association where the males provided meat and fish while the females collected seeds, tubers and insects; males defended the territory and hunted collectively, while females took care collectively of the offspring and obtained other food resources, which allowed the offspring, who needed longer care and greater protection, to survive long enough to become self-sufficient.

As we can see in the analysis of these three factors, it is possible that a cooperation strategy developed between the members of the population, where both males and females contributed to the survival of the progeny and increased their biological efficiency: *an efficient strategy as both the individual and collective interests met.*

Although bipedalism, food and helplessness of the newborn are the three factors I consider essential in the evolutionary process of our species, other factors also contributed to the appearance of an Evolutionary Stable Strategy (ESS)[13] at the population level, the emergence of a social order, and the origin of morality and law as a regulation mechanism of this social order. They are: *the collective upbringing*, the active sharing of food resources in *an exchange system of generalized reciprocity* and the selection of a mating system with a significant tendency to *monogamy.*

Concerning collective upbringing, we start with the supposition that hominid females and offspring were gregarious and remained together in an area that would both allow them to protect themselves against predators or competitors and to have access to water and various food sources. Moreover, if fertile females indeed had synchronized menstrual cycles, they would get pregnant and give birth simultaneously. With children of the same age, the probability of sharing breastfeeding and other parental care was greater, which made it feasible, as it is an efficient behaviour, for the females of the group to have helped each other during relatively long periods of time.

As biological efficiency increased, the number of children increased, and their presence generated increasingly specialized forms of exchange between adults and newborns. The symbolism of mother–child relations extended to relationships of friendship or appeasement – hugging, cuddling, kissing, cleaning – ensuring that affective behaviour leads to trust and security. This produces *attachment* to individuals of the same group, which is expressed in helping behaviour; it is sympathy beyond empathy that allows the individual to understand the needs of the other and to worry about the other. This way, a loop is established in which attachment makes cooperation and altruism possible. At the same time cooperative behaviour generates affective links. As stated by N. Acarín, this tendency to attachment is a good base on which to build a community life to protect each other, care for the offspring and undertake collective projects.[14]

It is in the context of a longer ontogenetic development and a collective upbringing that the adaptive cognitive-affective structure of individuals is produced that facilitates the coordination of their behaviour as members of a group, which in turn favours the survival of one and all.

The second factor that contributes to the emergence of an ESS at the population level is an inter-gender food exchange, and an active sharing that first might have appeared as largely intra-gender and that later generalized to the group. We had seen that males could have initially contributed in an individual way in the exchange of food for sex, as it is observed in bonobos and chimpanzees today. The more efficient team activity of hunting or scavenging, and then taking the meat to settlements to be chopped and shared could have increased the exchange of food for sex, the exchange of food between the two sexes, and finally the active sharing between individuals of the group, including the youngest who approached to beg.

The food exchange between the sexes and the later active sharing was possible because the males and females had their own survival strategies that led them to specialize in obtaining certain food resources. The breastfeeding requirements in females, a longer dependence of the offspring and the greater need for protection of the newborns could have led them to specialize in the collection of fruit, seeds, tubers, eggs and vegetables in the areas of the forest most protected from predators and competitors. On the other hand, males specialized in obtaining other kind of proteins and fat during their rounds of territorial defence.

As all resources were not available throughout the year and hunting has great probabilities of failure, subsistence depended on the sharing between males and females of the food available during the different seasons.[15] By sharing it within the group, the chances of starving decreased, thus increasing the biological efficiency of the population.

Moreover, the reiteration of this cooperative behaviour could have played an essential role in the development of reciprocal social obligations as linking behaviour at the population level. I speak of linking obligation and a sense of obligation because sharing between non-relatives is a cooperative, altruistic behaviour that runs the risk of not being met with a reciprocal response. If the behavioural pattern of reciprocal altruism was selected, it comes with emotional mechanisms that reinforce this behaviour and avoid or inhibit the behaviour of negative selfishness.

The third factor which led to an ESS was the selection of a mating system with a significant tendency to monogamy and with mechanisms to detect and control mating outside the couple (MOC), whether bidirectional or within the group.

This supposes that cooperation in males in the defence of the group territory, as well as the coordinated and cooperative obtaining of resources from hunting or scavenging, were only possible if masculine sexual rivalry did not imply direct confrontations of zero sum, in which only one or a few males had access to fertile females while the others did not. Where such cooperation was necessary, it was impossible to live in 'harems' of a male with many females, or in constant, direct confrontations for sexual access.

A hypothesis suggests that the reduction in the size of human canines may be due to decreased competition between males for females.[16] This reduction is already present in *A. Aphaeresis*, found by D. C. Johanson. Another hypothesis emphasizes the decreased dimorphism between males and females in relation with corporal size, which happened slowly in *H. ergaster / H. erectus*,[17] and which also suggests less direct confrontation for reproduction. Hidden ovulation could have also acted as an enhancer for monogamy. This factor prevents a dominant male from monopolizing all females during their highest fertility period because he simply would not know when it takes place.

For the males it was a more beneficial arrangement that several males in monogamous relationships became fathers at the same time than to try to monopolize a group of females in reproductive competition, or to have direct confrontations. For this arrangement improves the odds of collective defence, protection and care. As for the females, receiving cooperative protection by all males would be better than by just one or a few. Monogamy thus leads to beneficial synchronization. Given that the offspring are born more immature as time goes on, and thus need greater care and protection, the advantages of monogamy may become even larger.

In the evolution of our species, then, frequent mating and the hiding of visual signs of ovulation[18] can likely be associated with the change from polygamy to a mutual selection that offered trust and security to males and that generated affective links between males and females.[19] Ties of emotional closeness and love may increase the trust of males in the belief that the offspring is really theirs, and conflicts are thus reduced. 'If we introduce couples in such a society, social harmony can increase. Males can abandon the group for brief periods of time without losing the opportunity to be represented in the next generation. Parental care by males is possible, and the food can be shared', said Lovejoy in Johanson and Maitland.[20]

In short, a mating system with greater tendency to monogamy would increase male trust in paternity and would have favoured parental investment,[21] decreasing reproductive conflicts, and in turn increasing biological efficiency at individual and collective levels, thus widening genetic variability at the population level, which makes adaptation to new environmental pressures possible.

Moreover, I suggest that the selection of a mating system with greater tendency to monogamy was socially accepted because it contributed to the maintenance of a symmetric structure based on the preservation of non-zero sum interactions.

Although there is no unanimity in this regard, I think that the elements I have analysed allow me at least to defend this hypothesis of a monogamous mating system in our ancestors, although conditioned to the possibility of an alternative strategy:[22] opportunistic unfaithfulness for both sexes. I should make clear that the selection of a mating system in the evolutionary context of our ancestors does not mean that such a system is inflexible, for this is a variable conditioned by other circumstances, limitations or pressures, as it is reflected in the societies where we find polygamy or polyandry.

A Social Order, the Origin of Morality and the Law

The three factors that we have just analysed led to a cooperative and concerted integration of forces that produced a higher average probability of survival for the group members than the one each individual would have had separately.

Cooperative behaviour, group upbringing, the active sharing of food and the selection of the monogamous mating system were non-zero sum interactions which maintained or increased the biological efficiency at the population level even though the relative individual aptitude of some individuals in the group decreased.[23] This strategy was Evolutionarily Stable at the population level, for it reduced individual differences in reproductive success and material well-being; a mating system with a significant tendency to monogamy, the team-cooperative strategies and the food sharing are levelling behaviours that reduce the competition of individuals within the group.

As in the evolution of other social species, in which a systemic order emerges from recurrent interactions, in our species a pattern of social organization of symmetric structure emerged that was efficient because it conjoined the individual and collective interests, while maintaining a dynamic balance with regulation and control mechanisms over those behaviours that could destabilize social order.

While positive behaviours are reinforced through positive emotional dispositions such as feeling good when receiving approval or praise from others, negative behaviours are sanctioned, banned or inhibited through dissuasion, social sanction or punishment, which produce a sense of guilt, or as a result of the moral aggression by others in response to the offending behaviour.[24]

Since deception and fraud erode cooperative efforts for mutual benefit, cooperating members must control cheats and parasites. Positive selfishness proved to be effective as our ancestors had psychological tools that could guide or enhance cooperative actions while inhibiting those that were not. According to research by Gintis et al., human beings have a predisposition to cooperate with non-relatives, and also to identify and sanction social fraud, punish those who do not follow cooperation or reciprocity norms (independently of self-interest), based on a sense of equity reflected in the social sense of justice or injustice.[25]

In experiments using interactive games, M. A. Nowak and K. Sigmund concluded that a great part of human cooperation is based on moralistic emotions, such as the hate of cheats, or the sense of happiness and satisfaction when doing a good action, or the fear of being negatively evaluated and judged by others (reputation).[26] These emotions are felt even in front of individuals with whom no frequent interactions take place and of strangers.[27]

If during the hominization process there was sexual selection of the monogamous mating system, although with the possibility of an alternative strategy (MOC), this could provide an explanation of jealousy as an emotion intended to prevent being deceived by that person one needs to trust in order to produce and invest in offspring.[28] At the group level 'fair play' is positively valued, as is faithfulness, while 'adulterous' behaviours are negatively valued, since these could reduce the biological efficiency of the group by affecting individual cooperation for collective defence, food gathering and protection of the offspring. This system of values may have acted as a homeostatic regulation mechanism of the socially accepted mating system, which, while taking advantage of and reinforcing the links between couples, would also sanction infidelity.[29]

Efficient behaviours and their emotional mechanisms for reinforcement or sanction generate a sense of stability at the group level, as they are perceived and valued as the right behaviours by the members, as ethical 'oughts', and are thereafter viewed as social rules.[30] It is *as if* there had been a 'social contract'.

These behaviours rooted in relations of equality and justice in the form of customs (law) and feelings (morality) thus account for the origin of law and morality. Norms are then, the expression of a coordination phenomenon that acted as a mechanism of social balance as it prevented the accumulation of power in the hands of a single individual, while controlling cheats and spongers.

Norms reflect the emerging social order that becomes the collective *ethos*. In the social organization called *horde*,[31] social control is exerted by all: there is no central authority that imposes or punishes, it is done by the community as a whole on a basis of equality. There is no authority other than public opinion to control behaviour and to express moral judgements, which are not based on abstract concepts of good and evil: the horde considers undesirable, evil and the behaviour of individuals who violate the rules of reciprocity, reproduction and equality, for such behaviour adversely affects the social equilibrium.

Social Systems Today

Although these days we encounter different forms of organization systems, it is in the hunter-gathering *hordes*, typical of our ancestors, that we can confirm, to a certain extent, the origin and function of morality (*mores*) in our species. Studies by anthropologists describe these types of societies as having a structure of equality in which individuals are interdependent and reciprocity is the rule. Lacking a centre of power, individuals support and control one another by being integrated into a net of recurrent interactions in equality relations.

Customs possess two elements: a linking authority (group pressure) and interiority (obligation; either as an 'is' or a duty to choose from several behavioural options). Public opinion condemns those who do not follow the norms, and the fear of social punishment, of public exposure, if the transgression becomes public, is a strong dissuasive force for not breaking the tradition. I think that these customs become *law*, defined as the set of mandatory norms determining social relations imposed by the group; they are the expression of a phenomenon of coordination of certain behaviours that have the function of maintaining social equilibrium.[32]

This function of law as a regulation mechanism is the same in all current forms of society, even in highly stratified societies; the difference is that in these societies, there is an asymmetric structure and the function of law is now to maintain this asymmetry. It is in stratified societies of asymmetric structures where coercive norms advantageous to the rulers are formulated and imposed by those in power, where the law emerges as an instrument of social power. In these stratified social organizations where power (political–religious–economical) is in the hands of a few people, the norms define, give content and establish what is just or unjust, what is good or evil. At the same time, in these authoritarian organizations there are struggles to face dominants and change the system's structure. There are also reflections about equality, dignity and justice that challenge the content of authoritarian norms.

10 SYMMETRY AND EVOLUTION: A GENOMIC ANTAGONISM APPROACH

William M. Brown

> Parts which are homologous tend to vary in the same manner; and this is what might have been expected, for such parts are identical in form and structure during an early period of embryonic development, and are exposed in the egg or womb to similar coditions [*sic*]. The symmetry, in most kinds of animals, of the corresponding or homologous organs on the right and left sides of the body, is the simplest case in point; but this symmetry sometimes fails[1]
>
> – Charles Darwin, *The Variation of Animals and Plants under Domestication* (1868)

Darwin was one of the first biologists interested in the developmental failure of perfect symmetry, what is now known as fluctuating asymmetry (FA). FA is the random left–right size differences in bilateral traits designed by selection to be perfectly symmetrical. Thus at the population level genuine FA (as opposed to directional or antisymmetry, neither of which are the focus of this chapter) should be normally distributed with an asymmetry mean of zero. Arguably, L. Van Valen helped to facilitate the modern scientific study of fluctuating symmetry, despite a long history of its use.[2] FA is an indicator of underlying developmental instability which displays the inability of an organism to reach an adaptive end-point when experiencing adverse environments.[3] FA is often used as an indicator that structural features are developing under environmental and genetic stress; essentially FA may be an important example of a disruption in homeostasis.[4] Developmental instability can be measured in two ways: (a) as major developmental errors that occur as birth defects i.e. phenodeviants and (b) as subtle deviations from bilateral symmetry, i.e. FA, the latter of which is the focus of this chapter. Further, this chapter introduces a new genetic co-adaptation theory for developmental stability called *genomic antagonism reduction*. Simply stated, when intragenomic and intergenomic conflicts are minimized, individuals will be better at buffering ontogenetic stressors. Some previous findings are consistent with this proposition, and they will be reviewed here.

An array of studies in conservation biology have shown that asymmetries in bilateral morphological traits indicate underlying developmental instability.[5] It is assumed, due to considerable evidence, that low FA indicates good genotypic or phenotypic quality (i.e. good development).[6] FA covaries negatively with health and physical performance in a diverse array of species including humans, and appears to be a marker underlying the reproductive viability and health of a given phenotype.[7]

FA provides a 'window' with which to view how an organism has adapted to the environment and hence its ability to resist the harmful development perturbations.[8] A. P. Møller has reported that symmetrical individuals generally have better survival and mating success than their asymmetrical counterparts.[9] Perfect symmetry is the optimum for traits designed by selection to be perfectly symmetrical, the larger the FA the lower the developmental stability.[10] It is important to note that fluctuating asymmetries emerge with increasing exposure to a wide range of stressors such as pollutants, extreme temperatures and genetic perturbations.[11] For example, captive ornamental goldfish (*Carassius auratus*) and carp (*Cyprinus carpio*) showed significantly higher FA than wild samples in response to environmental stress such as overcrowding and lower water quality.[12] Further, J. A. Hódar found that leaf asymmetry in Holm oak (*Quercus ilex*) decreased in rainy sites compared to dry sites.[13] FA in house flies (*Musca domestica*) appears to be influenced by rearing conditions, showing higher FA at lower temperatures, and finally Møller et al. found that disruptive light regimes when rearing domesticated chickens (*Gallus gallus*) caused increased FA.[14]

Degree of symmetry can be shaped by artificial selection, such as selection for endurance wheel-running in mice and has been linked to natural selection and sexual selection outcomes.[15] In terms of sexual selection, studies across diverse species reveal that low asymmetry males tend to obtain more mates. In lekking black grouse (*Tetrao tetrix*) symmetry of their tarsi (the joint between the leg and foot) predicted copulation success.[16] FA in yellow dung flies (*Scathophaga stercoraria*) also accounted for decreased mating success.[17] J. P. Swaddle reports that FA is a negative predictor of hunting and mating success in the yellow dung fly (*Scathophaga stercoraria*).[18] Males in one class of British yellow dung fly (*Sepsis cynipsea*) that were more symmetrical in fore tibia length had more mates.[19] Paired male field crickets (*Gryllus campestris L*) were older, larger and more symmetrical.[20] Low wing length FA in house flies (*Musca domestica*) are found to have a higher mating success.[21]

Across diverse species, a large literature points to meaningful connections between FA, genotypic and phenotypic quality. It has been reported that FA is negatively correlated with attractiveness in humans.[22] The 'good genes' model of sexual selection assumes that due to the benefits associated with the selection of a healthy mate, i.e., enhanced offspring viability, mate preferences favour healthy individuals.[23] Along with this, studies have found a relationship between attractiveness and facial symmetry.[24] Likewise, lower FA is related to an array of human sexual behaviours, such as increasing number of sex partners, younger age of first sexual contact and sexual contact outside a stable relationship.[25] Furthermore, FA is negatively associated with the attractiveness of the human vocalizations, a sexually dimorphic feature in humans.[26]

Examinations of physiological and physical health correlates of FA have shown an array of theoretically consistent associations.[27] For example, sex-typical body size is associated with low FA.[28] Male testosterone levels (an important correlate of muscle development) and athleticism are both positively associated with attractiveness possible due to underlying developmental stability.[29] Indeed, one would expect that greater body mass in men (within the normal range of healthy variation) would be negatively correlated with FA.[30]

FA is also associated with diverse outcomes associated with central nervous system development in humans. For example, R. M. Malina and P. H. Buschang tested a group of normal and retarded males and found that FA was significantly greater among mentally retarded individuals compared to normal subjects.[31] This suggests a possible connection between developmental instability and central nervous system functioning. R. J. Thoma et al. found a negative correlation between FA and neural speed as measured by magnetoencephalographic (MEG) dipole latencies during a sensory-motor integration task (i.e., index finger response to a visual stimulus).[32] Increased FA was a predictor of slower neural-processing across all stages of the visual-motor task in a sample of men. There are other diverse psychological correlates with asymmetry. For example, patients with schizophrenia have increased brain asymmetries.[33] Finally, S. M. Martin et al. found a positive relationship between FA and depression in men.[34]

Increasingly there are negative associations between FA and locomotory efficiency (a commonly used proxy for physical health). Areas relating to mechanical efficiency across species have shown that low FA is associated with increased performance.[35] For example, wing asymmetry in wild-caught adult European starlings (*Sturnus vulgaris*) revealed that wing asymmetry is negatively related to flight performance whereby investigations indicate that asymmetry of both wings and tails decreases mobility and swiftness.[36] J. Martin and P. López found in male lizards (*Psammodromus algirus*) that femur length FA resulted in decreased escape speed and ability.[37] In human studies, low FA was associated with higher running ability rankings.[38] More recently, D. Longman, J. T. Stock and C. J. Wells found that FA was a negative predictor of rowing performance.[39] These associations could be due to anatomical impairments or possibly neuromuscular coordination deficits. For example, of anatomical impairments, lower back pain sufferers have greater asymmetry in pelvis, ulnar length and bistyloid breadth.[40] The positive association between dance ability and symmetry in a healthy sample suggests that motor movements and neuromuscular coordination may also reflect degree of underlying developmental instability.[41]

Weaknesses and Future Research

Despite the FA literature revealing significant associations with biologically important outcomes across diverse species, including humans, the field is not without controversy. Most human FA studies have a similar rationale (i.e., researchers expect that since high FA reflects an inability to reach an adaptive end-point, then it must correlate with all things bad). However, there are notable exceptions to this pattern worth considering. W. M. Brown and C. Moore predicted that since high FA individuals may be vulnerable to extrapair copulations, then they may benefit from increased mate guarding behaviours via a facultative use of romantic jealousy.[42] Indeed this is what Brown and Moore found, and it highlights the importance that if an organism is developing under genomic or environmental stress, it may employ alternative tactics to cope with the costs of high FA (i.e., 'doing the best in a bad situation').[43]

Another misconception is the idea that since low FA is an indicator of 'good genes', this asymmetry level must be fixed at birth. Note that 'good genes' has largely been used as short-hand for good development (where genes only play a part among other factors). Regardless of this terminological issue (i.e., accuracy of the good genes metaphor), it is important to note that FA is not fixed at birth: it changes over the course of development and there is mounting evidence that there is a yet to be discovered developmental homeostatic mechanism of compensatory growth.[44] Specifically, there is solid evidence that across diverse taxa, when one side is larger than the other, growth slows for the larger side and speeds up on the smaller – essentially correcting the previous subtle asymmetry.

There are mixed findings in the study of FA (although presumably no more than any other area of evolutionary ecology – see A. P. Møller and R. Thornhill).[45] These mixed findings may reflect the weakness of FA as a general measure of latent developmental instability, problems with measurement error or a combination of both. It has been argued that the effect size and heritability for FA is rather low, but not necessarily lower than any other trait studied in ecology. Nonetheless, developmental stability holds a special place in evolutionary biology since it rests at the fundamental core of how genes and the environment are intertwined during the ontogeny of phenotypes. Early FA researchers would refer to low FA individuals as displaying fitness indicators or perhaps displaying their 'good genes' to the opposite sex. Generally such short-hand has subsided within the main literature as few studies have actually investigated the associations between FA and molecular loci. This has now begun to change with the increasing availability of molecular and cellular assays of healthy development. One example is the study of FA and heat shock proteins.

Heat shock protein (*Hsp*) activity under some circumstances may be a better assay of developmental buffering capacity than FA.[46] *Hsp*s are an evolutionarily-conserved protein class important for the regulation of other proteins when exposed to stress. They are believed to be an important aspect of an organism's developmental buffering capacity. For example, in one study of *Drosophila*, *Hsp90* chaperone buffering capacity was not correlated with FA, but in another study, K. H. Takahashi et al. measured the activity of four different *Hsp* genes (*Hsp22, Hsp67Ba, Hsp67Bb, Hsp67Bc*) and found the predicted association (i.e., deletions of relevant *Hsps* led to an increase in *Drosophila* bristle FA).[47] So far there have been no studies testing the link between human FA and *Hsp* buffering capacity. However, given the recently reported positive association between human FA and oxidative stress – a correlate of *Hsp* response – by S. W. Gangestad, L. A. Merriman and M. Thompson, developing multiple assays of developmental buffering capacities at different levels of biological organisation, may be the best way forward for future research due to problems with effect size and measurement error in FA studies.[48]

Conflicting Theory in the Study of Symmetry

In some ways the study of FA has been part of the scientific pursuit to find an accurate assay for the developmental integrity of organisms (i.e., the unity of their component parts). In this sense the big questions in the scientific study of symmetry have not really diverged much from Darwin. How do our developmental integrity genes (presuming such entities exist) regulate symmetry in developing organisms facing environmental and genetic stressors? This outstanding question may be resolved in part by remembering that bilateral traits have one genotype, so any differences between sides are an epigenetic phenomenon. Modern epigenetics is defined as heritable changes in gene expression that do not change the underlying DNA sequence. Bilateral traits are analogous to monozygotic twins (i.e., they share the same genotype). Specifically, when size and shape differences emerge between the left and right sides, they may be caused by epigenetic forces that are only recently being addressed via molecular study (e.g., investigating mechanisms such as DNA methylation, histone modifications and microRNAs).

The remainder of this chapter will present a new theory of developmental stability specifically designed to address existing theoretical controversies surrounding the generality of the relationship between heterozygosity and FA. Heterozygosity is when two alleles on each strand differ from one another. There are two rival approaches regarding the emergence of developmental instability in nature each bolstered by contradictory empirical findings regarding the FA-heterozygosity association: (1) the co-adapted gene complex hypothesis (simply stated heterozygosity and FA are predicted to be positively related); and (2) the heterozygous advantage hypothesis (simply stated FA and heterozygosity should be negatively correlated).[49] It is proposed in this chapter that a reduction of genomic antagonisms mediates the contradictory associations found between FA and heterozygosity.

Developmental Stability: A Genomic Antagonism Approach

Most scholars acquainted with the study of evolution are aware of Richard Dawkins's idea that genes are selfish replicators; however, as importantly pointed out by J. E. Strassmann, D. C. Queller, J. C. Avise and F. J. Ayala, selfish genetic elements behave in particularly selfish ways, as their replication success often supplants the interests of the organism itself.[50] These ultra-parasitic elements have unique biochemical pathways which can be in conflict with other loci in the genome (Figure 10.1 illustrates different types of selfish genetic elements).

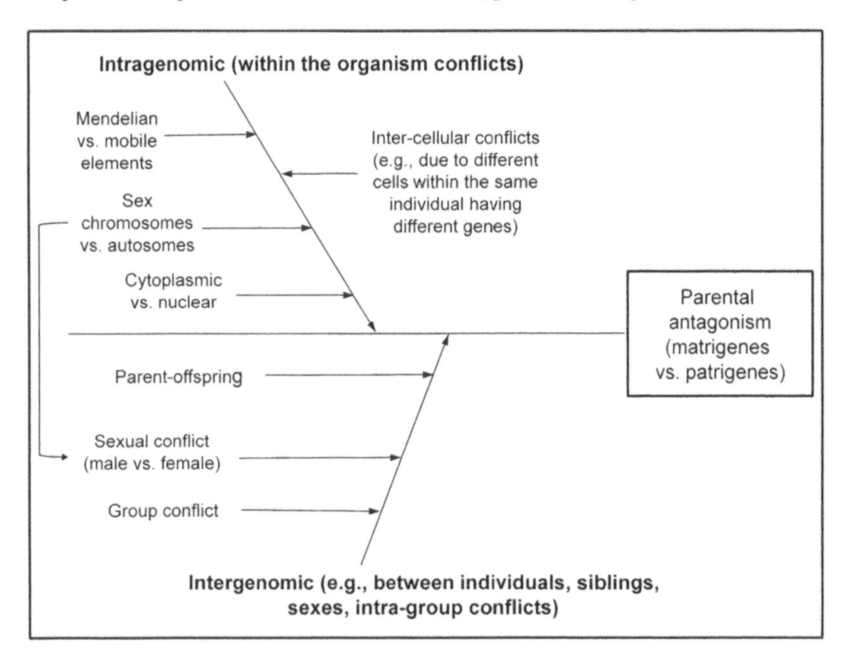

Figure 10.1: Types of intragenomic and intergenomic conflicts, modified from J. H. Werren, 'Selfish Genetic Elements, Genetic Conflict, and Evolutionary Innovation', *Proceedings of the National Academy of Sciences of the United States of America*, 108 (2011), pp. 10863–70.

J. H. Werren hypothesized that selfish genetic elements are evolutionarily maintained by their selfish behaviour, but the new chunks of DNA left throughout genomes can sometimes be co-opted, domesticated, or modified to cause some beneficial effect to the organism.[51] Figure 10.1 highlights the importance of parental genome conflicts, which mediated in part by epigenetic molecular tags of shared recent ancestry.

Importantly we see that organisms in principle are not cohesive wholes. Within genomes there are sometimes conflicts. Intragenomic contest evolution or sexual antagonistic effects exist, also parental antagonisms between parentally-derived alleles within offspring. The idea that there is a quantifiable developmental integrity must deal with the presence of selfish genetic elements. Interestingly there has been little work on this for the study of symmetry (however there are notable exceptions to be discussed below).

Despite the diversity of selfish genetic elements in Figure 10.1, there are some general observations regarding their frequency in nature.[52] For example, the diversity of selfish genetic elements in a species is positively correlated with outbreeding rate. Indeed it is expected that sexual reproduction enhances the spread of selfish genetic elements, while inbreeding decreases it.[53] Heterozygosity decreases upon inbreeding and opportunities for transmission between genomes decreases in a population. Thus among inbred populations, selfish genetic elements have an increased chance of being paired with their homologue and the frequency of selfish behaviour would decline impairing spread of the element. Among inbred strains of yeast (*Saccharoyces cerevisiae*) a homing endonuclease gene – considered to be a selfish genetic element – there was no significant increase in frequency of the selfish genetic element compared to outcrossed samples.[54] The work of M. R. Goddard et al. provides convincing experimental evidence for the effect of inbreeding to dampen the spread of a selfish genetic element.[55] Importantly, this work provides evidence that the host mating system plays a critical role in the population dynamics of a selfish genetic element.

Based on the above findings, the idea presented in Table 10.1 is that the contradictory correlations between FA and level of heterozygosity may depend on the presence of selfish genetic elements within the genome. Specifically, when antagonisms are low (e.g., among some plants and ectotherms), a negative association between FA and heterozygosity will emerge; however the reverse will be true when genetic antagonisms are high (e.g., among social living organisms). In the latter case, social-living endotherms may benefit from homozygosity as it could weaken the effects of genetic antagonisms (e.g., conflict between parental genomes).

Table 10.1: Predictions derived from genomic antagonism reduction theory regarding FA outcomes by levels of heterozygosity and genomic conflicts.

		HETEROZYGOSITY	
		LOW	HIGH
ANTAGONISM	LOW	↑ FA	↓FA
	HIGH	↓FA	↑ FA

Based on genomic antagonism reduction theory for developmental stability, a possible solution emerges for the contradictions between the genomic co-adaptation versus heterozygote advantage approaches to developmental stability (i.e., the latter predicts that low FA organisms will have high heterozygosity, while the former predicts the reverse pattern).

These contradictions may depend upon the degree to which social relations affect the spread of selfish genetic elements and underlying intragenomic antagonisms. Social conflicts with kin are caused by distinct species and life history factors (i.e., live bearing, heat sharing, internal gestation, postnatal care, relatedness asymmetries due to dispersal and mating systems etc.). One way to reconceptualize this problem is the development of a theory that takes into account the degree of intragenomic antagonisms for a given species or individual. Essentially, when testing heterozygosity-FA relationships, one should consider species' natural and life history background for the degree of genomic conflict. The cross-species FA-heterozygosity associations reported by L. A. Vøllestad, K. Hindar and A. P. Møller indicate a pattern largely consistent with the genomic antagonism hypothesis, which is explored further in the following paragraphs.[56]

Vøllestad, Hindar and Møller found the predicted negative association between FA and heterozygosity (for within-population studies of ectothermic species only).[57] One reason for this association not emerging for endotherms may be due to the degree of parental antagonism among internally gestating and heat-producing species. Indeed it appears that in Vøllestad, Hindar and Møller, among organisms with conditions ripe for genetic conflict (i.e., internal gestation, heat sharing in social contexts) the negative association between FA-heterozygosity reduces.[58] One may ask: why would heat production and sharing be important for the development of underlying genomic conflict? The driver is when heat is shared among littermates or others with varying degrees of genetic relatedness between heat givers and receivers. Essentially, once endothermia emerged evolutionarily, conflicts over heat generation and reception may occur. When paternal genomes benefit from withholding heat from others, at a cost to the maternal genome, maternal counter-adaptations are expected to evolve (e.g., sharing heat among litter mates or asymmetrically related collateral kin).

Endothermia is a classic example of developmental stability. That is, altering internal mechanisms of heat generation is a homeostatic response to environmental disturbance. In the case of nonshivering thermogenesis, there is now empirical evidence that maternal genomes are involved with heat sharing among litter mates, while paternal genomes are more selfish.[59] Specifically, heat generated by huddling littermates is a collective good, and the empirical evidence suggests that there are two paternally expressed loci for reducing heat and one maternal loci involved with increasing heat to share with huddling littermates.[60] It stands to reason that endotherms are more likely to be engaged in intragenomic conflicts due to heat social exchange between rivals (e.g., thermoregulatory social conflicts mediated by brown fat or other mechanisms). Endotherms are unintentional temperature altruists. Once endothermia emerged evolutionarily, the context was set for conflicts over sharing one's heat with others (e.g., huddling littermates). This hypothesis may seem far-fetched; however, there is solid evidence that some endotherms with brown adipose tissue show the traces of an ancient intragenomic conflict of social origin. Specifically, genes for storing heat for oneself are paternally expressed, while genes for sharing heat are maternally expressed. Indeed simply being an endotherm (with brown fat), could be a rough indicator of genomic antagonism. It is also important to note the existence of behavioural-induced endothermia among ectotherms (known as facultative endothermy in insects, e.g., honeybees).[61] Indeed, some insect species are capable of maintaining higher than ambient abdominal temperatures using exercise behaviours. So, why do they? One hypothesis is to alter investment in kin and this will produce intragenomic conflicts over the muscle adaptations producing the behaviour. In these social ectotherms with facultative endothermy we would not expect the negative association between FA and heterozygosity. Indeed, in Vøllestad, Hindar and Møller's cross-species review report there is a negative association between FA-heterozygosity among ectotherms, not ectotherms.[62] Eusocial insects are facultative endotherms (i.e., normally ectotherms unless they create heat for the benefit of the colony) and may not fit the ectotherm pattern. As expected, the average association between FA and heterozygosity ($M = 0.003$, $SD = 0.095$) among fourteen studies of honeybees *Apis mellifera* (over 4,000 subjects), diverged from the ectotherm pattern (i.e., a significant negative association). That is, a one-sample t-test with a test-value of '0' revealed little association in honeybees between FA and heterozygosity: $t\ (14) = 0.108$, $p = 0.916$. A similar pattern emerged in ants (*Iridomyrmex humilis*), although to my knowledge there is no evidence of facultative endothermy.[63] *Iridomyrmex humilis* do have multiple queen colonies, which would in theory increase genomic antagonisms. A genomic antagonism reduction approach would suggest that in the case of high genomic antagonisms, hetrozygosity may be more detrimental to developmental stability (i.e., higher FA) relative to homozygosity. Further work is needed on eusocial *Hymenoptera* to determine if they do in fact diverge from the negative relationship between heterozygosity and FA found in many ectotherms.

Symmetry, Selfish Genetic Elements and Conflict Resolution

One of the key premises of this chapter is that selfish genetic element elements will increase developmental instability. There is some evidence for this association. For example, L. J. Leamy, S. Meagher, S. Taylor, L. Carroll and W. K. Potts found that a selfish genetic element called the t-allele was associated with increased FA in house mice.[64] However, in a study of yellow-necked mice (*Apodemus flavicollis*) results were less straightforward.[65] Specifically, trait size variability (but not FA) was positively related to increasing B chromosome number (a well-studied selfish genetic element). Interestingly, the levels of the FA followed the changes in frequency of subjects with B chromosomes at the beginning of the mating season. Granted, it could be that unknown ecological factors influenced a disruption in developmental homeostasis and the frequency of animals with B chromosomes.[66]

Interestingly, J. A. Zeh and D. W. Zeh hypothesized that organisms may use mating strategies to avoid selfish genetic elements.[67] Specifically, when a selfish genetic element reduces the competitive ability of sperm in those that are carriers, females may use polyandry as a strategy to avoid carrier males, as the fitness of their progeny would be more likely to be compromised if they inherit the selfish genetic element. Thus Zeh and Zeh theorized that polyandrous mating may be a tactic to reduce effects of genomic incompatibility caused by selfish genetic elements.[68] The antagonism reduction theory of developmental stability argues that when intragenomic antagonisms are reduced (e.g., low relatedness asymmetries, reductions in selfish genetic elements due to lower heterozygosity) individual organisms will have greater developmental stability. To indirectly test this hypothesis, Cuervo and Møller's cross-species avian FA data was categorized by whether a species is polyandrous, socially monogramous, lekking or polygynous. D. Hasselquist and P. W. Sherman have shown that chicks from extrapair matings are twice as common in socially monogamous compared to polygynous passerines.[69] If polyandry is a strategy to avoid genetic incompatibility from selfish genetic elements as hypothesized by Zeh and Zeh, it may have effects on developmental stability of offspring.[70]

In seventy species of passerines, Cuervo and Møller provided useful data on FA differences by species, sex and mating system.[71] Species' mating systems were classified based on previously published findings: 'socially monogamous' when a male and a female assortated for reproduction; 'polygynous' when approximately 5 per cent of the males were reproductively associated with more than one female; polyandrous when least 5 per cent of the females were associated with more than a single male for reproduction, and lekking when males aggregated at communal display grounds where females arrived to make their mate choice. Presuming that multiple paternity decreases the costs associated with selfish genetic elements, one would expect that socially monogamous/polyandrous species will have lower FA than polygynous/lekking species.[72] Such a finding could indicate that multiple paternity may increase offspring developmental stability via the reduction in incompatible matings due to selfish genetic elements. To test these hypotheses the published data from Cuervo and Møller were investigated.[73]

It was predicted that socially monogamous/polyandrous species of passerines would have reduced FA compared to non-monogamous species (e.g., lekking or polygynous). Counter-intuitively perhaps (as one could have predicted that where FA is sexually selected for low values then directional selection would have caused reduced FA in such species – the so-called Lek paradox), non-ornamental FA was significantly lower among monogamous/polygamous ($M = 0.53$; $SD = 0.28$, $n = 82$) compared to lekking and polygynous species ($M = 0.89$; $SD = 0.98$, $n = 53$): $t\,(57.55) = 2.59$, $p = 0.01$ (Figure 10.1). Note that between-group FA variances also significantly differed (Levene's test $p < 0.01$).

In contrast to non-ornamental FA (where the sexes did not significantly differ > 0.60), there were sex differences in ornamental FA, with males having higher FA ($M = 4.68$, $SD = 7.39$, $n = 61$) than females ($M = 2.20$, $SD = 3.00$, $n = 41$): $t\,(85.35) = 2.35$, $p < 0.05$. Males had significantly (Levene's test $p < 0.05$) greater FA variance than females. Finally, ornamental FA was not statistically different depending upon mating system ($p > 0.40$). This is indirect evidence for the genomic antagonism reduction theory of developmental stability and Zeh and Zeh's proposal that polyandrous matings reduce genomic incompatibly.[74] More generally it suggests that FA varies by mating system in passerines.

Summary

The premise of this chapter has been that when genomic antagonisms are minimized, developmental stability is expected to increase. This is a preliminary hypothesis and this review has just begun to unravel the complex processes and interactions between the multiple levels of organization responsible for genomic conflict resolution and its possible role in developmental stability. The main findings were that: (1) the negative associations between heterozygosity-FA may be mediated by the degree of genomic antagonism; specifically, species with reduced genomic antagonisms will receive developmental stability benefits from increasing heterozygosity; and (2) among passerines, polyandrous matings and mating systems may yield benefits for developmental stability.

There appears to be increasing evidence suggesting that positive developmental outcomes are associated with the reduction of genomic antagonisms. M. Kawahara and T. Kono have provided empirical support for the general idea even though they did not measure morphological asymmetries.[75] Specifically, in newborn mouse pups that were genetically manipulated in two regions – the imprinting centres of *Igf2-H19* and *Dlk1-Gtl2* – Kawahara and Kono found that that genetically manipulated mice (reduction in intragenomic conflict) lived 186 days longer than controls.[76] This extension of the longevity of progeny from bi-maternal genotypes in mice is consistent with models based on sex-specific selection of reproductive strategies causing differential effects on mortality.[77] It remains to be seen if such experimental manipulations of degree of genomic conflict reduce offspring FA. It is important to note that some of these mechanisms could well be mediated by behavioural adaptations. For example, mate choice and dispersal patterns. Future work needs to study how behavioural adaptations influence the degree of intragenomic conflict and optimize organismal developmental stability.

Conclusions

Organisms are integrated wholes but divided in principle.[78] This may appear contradictory for physiologists and organismal biologists not familiar with the concepts of selfish genetic elements (i.e., elements that harm the host for their own transmission advantage), parental (kinship theory of genomic imprinting) and sexual antagonisms (e.g, sexually antagonistic coevolution). Incorporating such genomic antagonisms into our current views of developmental integrity of organisms could well help resolve debates surrounding the associations (and lack thereof) between FA and fitness outcomes. Essentially, like most work in evolutionary ecology, the natural and life history of organisms shifts the trade-offs stable developers face. Essentially 'good genes' in one context may be 'bad genes' in another.

In conclusion we need to further explore the source and mechanisms mediating antagonism resolution, especially in terms of developmental integrity. Upon the discovery of the mechanisms mediating genomic conflict reduction, we may well explain two puzzles simultaneously: (1) contradictory findings regarding FA-developmental stability; and (2) why some organisms thrive with an abundance of selfish genetic elements and others do not.[79]

11 BEAUTY, BACTERIA AND THE FAUSTIAN BARGAIN

Victor S. Johnston

Introduction

According to Stephen J. Gould, we live in the age of bacteria.[1] Indeed since life began on this planet more than three billion years ago it has always been, and probably always will be, the age of bacteria. Their numbers are staggering, now estimated to be more than five million trillion trillion, far exceeding the population of all other independent living organisms.[2] They are in us, on us and all around us. On a cellular count, 'our body' is mostly bacteria. Only about 10 per cent of our cells contain human DNA while the rest are bacterial cells that inhabit our intestines, our mouth, or eyes, our nose, our hair and every other exposed crevice of our body. They are both our best friends and our worst enemies, and we are only beginning to understand the paramount role they have played in unconsciously shaping human sexuality and mate choice.

Parasites: Friends and Foes

About 2.8 billion years ago, long before the evolution of plants, it was bacteria (cyanobacteria) that oxygenated the earth's early atmosphere, turned the sky blue and converted inert atmospheric nitrogen into the fixated form required by all plants and animals. By appropriating these two magical chemistries, photosynthesis and nitrogen fixation, the earth's green plants became the primary producers in the food chain and the dominant energy supply on which all animal life depends. Humanity could not survive without these self-replicating, green energy, solar collectors, nor could it survive without the bacterial mediated digestion, disposal and recycling of human waste. We certainly owe a great deal to our dearest friends, but they are also one of our deadliest enemies.

In 1896 the renowned physician Sir William Osler stated that 'Humanity has but three great enemies: fever, famine, and war; of these by far the greatest, by far the most terrible, is fever.'[3] Fever is the body's response to bacterial, and sometimes viral, infections. Over the course of human history such pathogenic parasites have been the primary cause of human morbidity and mortality. In the sixth century the great plague, caused by the bacterium *Yersinia pestis*, killed more that 100 million people as it swept across the Middle East, Europe and Asia, and a second pandemic in the fourteenth century killed almost half the population of Europe. Typhoid, smallpox, cholera and measles have all taken their toll of human life. Even today, in the era of multiple antibiotics and antiviral agents, *Mycobacterium Tuberculosis* kills one human every twenty-four seconds, deaths due to HIV/AIDS have reached two million per year and the malaria parasite, blamed for killing about half of all humans who ever lived on this planet, still takes more than 800,000 lives each year. Most biologists now believe that it was such persistent, changing and deadly threats of invasion by parasites that eventually led to the evolution of sexual reproduction.[4]

The Cost of Sex

As George Williams argued, sex is biologically expensive.[5] In a sexually reproducing species, only half of a parent's genes are passed on to its offspring. In contrast, if the same parent were to reproduce asexually, by cloning or parthenogenesis, the offspring would receive all of its genes. From any gene's perspective sexual reproduction reduces its probability of survival by one half, so it is difficult to explain why it would ever be favoured over asexual reproduction. The cost of sex is even greater when one considers that it breaks up successful genotypes and involves such additional costs as finding a mate and rearing non-cloned, compared to cloned, offspring.[6] Given these circumstances, for sex to evolve there must be a significant biological benefit that exceeds these combined costs.

There is now substantial evidence that the genetic variance generated by sexual reproduction endows multicellular organisms with a resistance to the multitude of threats posed by rapidly evolving parasitic microorganisms.[7] Survival depends upon changing the cellular binding sites used by parasites and generating novel variance in the immune response; failure to change is a prescription for extinction. This explanation has been dubbed the Red Queen Hypothesis after a famous statement by the red queen in Lewis Carroll's book, 'Through the Looking-Glass'. When the queen stated, 'Now, here, you see, it takes all the running you can do, to keep in the same place', she captured the essence of the perpetual host–parasite arms race.[8]

Sexual reproduction does not necessarily involve two different sexes (males and females) who produce gametes (sperm and ova) that differ in size (anisogamy). Offspring could be produced by the fusion of two small equally sized gametes (isogamy) but this reproductive strategy may be unstable in multicellular organisms because natural selection would favour individuals who produce larger gametes with sufficient nutrient to support early cell divisions. When such large gametes are abundant, however, the production of small mobile gametes (sperm) that can seek out and fertilize large ova becomes a viable second strategy. Under these circumstances females, defined as the macrogamete producers, benefit from guarding their large investment by both internal fertilization and selective mate choice. It appears that the omnipresent menace of pathogenic parasites has driven many species into this intricate and costly form of sexual reproduction and it is now the dominant arrangement found in all birds and mammals. For humans, this outcome has far reaching consequences for their physical appearance and mate choice decisions.

Female Displays of Fitness

For most human beings mate selection is a complex multifaceted decision that is of paramount importance for their future happiness. It is the stuff of great novels and plays, comedies and tragedies, filled with pleasure or fraught with pain. Given the high stakes involved in this life-changing decision it is both strange and remarkable that an apparently trivial attribute, physical beauty, should play such a prominent role.[9] Our common sense tells us that beauty is only a personal whim (beauty is in the eye of the beholder) or a frivolous quality (beauty is only skin deep) that we should ignore (don't judge a book by its cover). Despite these admonitions, however, beauty is not ignored. Indeed, the enhancement of beauty has spawned several multi-billion dollar international businesses: the cosmetics industry, cosmetic surgery and advertising. It appears that individuals are willing to invest a great deal of time, money and effort in the pursuit of this ephemeral attribute. So perhaps there is more to beauty than meets the eye.

In the early nineties, a series of experiments suggested that the average female face in a population is the most attractive, and the authors proposed that this preference could have a biological function.[10] They argued that the average could be attractive because it possesses evolved facial features that are or were adaptive for survival in the local environment, such as long noses with thin nostrils in cold dry climates and short noses with wide nostrils in warm humid environments. Clearly offspring would benefit if individuals were attracted to faces displaying such adaptive traits. However, the averaging procedures used in these studies were problematic.[11] Subsequent experiments revealed that although the average face is attractive the most attractive face differs from the average in a systematic manner.[12] Specifically, female faces with small narrow chins, large eyes, and fuller lips are rated highest in beauty across many different cultures.[13] The significance of these features appears to lie in their hormonal origin.

Young boys and girls enter puberty with almost identical proportions of muscle, fat and bone, but rapidly mature into reproductive adults with very different body shapes and compositions. For the most part these anatomical changes are a product of steroid hormone levels. Under the influence of estrogens a young woman gains about thirty-five pounds of fat, changing the shape of her breasts, hips and lips. In contrast, a young man acquires about one and half times as much muscle and bone mass, regulated by the complex action of androgens (and aromatized androgens) acting both directly and indirectly (via release of growth hormone) on bone and muscle tissues.[14] As a consequence, the adult male's face has a longer and broader lower jaw than that of a female and the growth of brow ridges results in more sunken narrow eyes.

Viewed from this hormonal perspective, the most attractive female faces are displaying physical features that signify higher levels of pubertal estrogens (full lips) and lower levels of androgen exposure (short narrow lower jaw and large eyes) than average females. This combination of hormones also appears to be responsible for the low 0.7 waist to hip ratio that has been found to be a universally attractive feature of female bodies, and associated with physical health and high fertility.[15] In the absence of contraception, female fertility reaches its maximum in the mid-twenties, declines by about 20 per cent in the mid-thirties, and then falls precipitously by a further 60 per cent during the forties.[16] The thinning of a female's lips parallels these steep declines in fertility and it is not uncommon for females to use lipstick or collagen injections for maintaining or enhancing their facial attractiveness. Taken together, these observations suggest that female beauty depends upon specific highly visible hormonal markers that signal higher than average fertility.

Fitness Monitors

In humans, such fertility displays would be ineffective if they did not evoke an emotional response in the brains of male admirers. This hypothesis can be evaluated by recording event related potentials (ERPs) from the brains of human volunteers exposed to brief presentations of facial pictures. Prior studies have established that the amplitude of the third positive wave (P_3) of ERPs increases systematically with the emotional value of a stimulus to the observer.[17] To study the emotional value of facial features ERPs have been recorded from males exposed to a random sequence of male and female facial images designed to systematically manipulate the size and shape of facial features.[18] The results reveal that for female faces, but not male faces, the P_3 amplitude is highly correlated with males' beauty ratings of females, and the largest P_3 response is evoked by female faces displaying full lips and a short narrow chin, the feature combination postulated to be an index of high fertility. It appears that male brains are exquisitely sensitive to these hormonal markers and respond to such cues within 500 milliseconds: the latency of the P_3 component. In the real world this implies that a man could probably assess the beauty of a woman's face in a single glance across a crowded room!

Sexual Selection

Hormone mediated fertility signals such as chest blisters, labial swelling or face reddening are common in other female primates.[19] Females use such signals to attract males and those attracted by such cues enjoy enhanced reproductive opportunities. Nevertheless, in species that reproduce by internal fertilization, female choice is the critical factor that controls reproductive success (RS). Compared to males, females make a large initial investment in their offspring in the form of a large ovum. After fertilization this investment is amplified by a period of incubation or pregnancy, followed by lactation in the case of mammals. However, because internal fertilization guarantees that her genes are in the offspring, this high parental investment normally assures high RS. Ultimately, however, the degree of this success depends upon her mate choice; selecting a male with 'good genes' and/or the resources to ensure offspring survival.

Elaborate mating dances, melodic songs and displays of brightly coloured iridescent tails, are some of the lures employed by males to seduce members of the opposite sex. The function and effectiveness of these testosterone-mediated signals has been studied using experimental 'plastic surgery'. For example, extending the tail feathers of widowbirds produces super-tailed males who enjoy more reproductive opportunities than their average-tailed competitors.[20] In the absence of human interventions, females who mate with attractive males appear to reap real biological benefits. Peacocks with large tails have higher survival rates but, more importantly, this enhanced survival is evident in his mate's offspring.[21] From a peahen's perspective it appears that a male's beauty is not just an empty promise; she can enhance her RS by selecting males who exhibit these extraordinary secondary sexual characteristics. Of course all such elaborate displays require significant energy to produce and flaunt so they automatically attest to the physical health of a male suitor, but the demonstrated benefits to an attractive male's offspring suggests that 'good genes' may also be involved.

Male Displays of Fitness

Image processing software has allowed experimenters to systematically manipulate the degree of testosterone markers on the facial images of human males.[22] These studies have revealed that females prefer male faces that are more masculine than the average male and this preference becomes more extreme at ovulation or when selecting the face of a short-term mate, compared to a long-term mate – occasions when there is either a higher probability of conception or little expectation of resources other than 'good genes'.[23] The relationship between masculine secondary sexual traits and 'good genes' is supported by studies of fluctuating asymmetry (FA). FA is the measured deviation from perfect bilateral symmetry of those physical traits for which signed differences between the left and right sides have a mean of zero over the population.[24] Across many species, including humans, males with low FAs enjoy better health and more mating success than asymmetrical males.[25] Such asymmetries can be caused by pathogenic parasites or other insults encountered during the course of development, so low FA is believed to be a valid index of a competent immune system.[26] Since there is a significant positive correlation between low FA and facial masculinity in human males, facial testosterone markers can serve as a visible proxy for good genes.[27] If this is the case then the host–parasite arms race, the raison d'être for sexual reproduction, also influences human mate choice decisions.

Adaptive Illusions

Humans do not see or hear their deadliest enemies, but they are aware of their presence from the smell of rancid meat, the taste of sour milk or the noxious odour of waste products. It is not a fortunate accident, however, that such unpleasant feelings are evoked by contaminated food; it is an essential part of our biological design. The obnoxious smell of rotten eggs is not a property that belongs to the hydrogen sulphide molecule; it is an evolved emergent property of the human brain. Hydrogen sulphide might smell wonderful to a dung beetle! Sugar is not sweet, nor toxins bitter, for all such pleasant or unpleasant feelings have evolved as discriminations between a diverse range of potential threats or benefits to our RS.[28] Individuals need not be aware of the connection between a foul odour and bacterial contamination, between pain and tissue damage or between the sweet taste of sugar and the production of ATP, for natural selection has already forged the relationship between our feelings and our gene survival; a design made evident by the intensity of sexual orgasms experienced at the imminent moment of our reproductive success. From this perspective beauty is like sugar, we like it without any cognitive awareness of its biological value. But like the sweetness of sugar it only exists as a positive feeling generated by a human viewer's brain in response to environmental cues that indicate a potential benefit to his/her reproductive success; it is, in essence, an evolved adaptive illusion of the human mind.[29]

Conclusion: A Faustian Bargain?

Asexual organisms are potentially immortal, but sexually reproducing organisms are destined to die. The hidden cost of sex is eventual death, and we humans are painfully aware of our future demise. We fear serious illness and death, but such feelings are also part of our biological design. The world inside and outside our skin simply obeys the unfeeling laws of physics and chemistry and there is no 'good' or 'bad' inherent in the execution of such laws. That is, if a bullet pierces the heart of a human child the laws of physics are not broken and the chemistry of the child's death is neither a 'good' chemistry nor a 'bad' chemistry, it is just ordinary chemistry. It is the evolved evaluative feelings generated within the brains of human beings that add meaning to such otherwise meaningless happenings; the sting of death is in the pain, the sorrow or the fear, not the physics. Without such emergent feelings, evolved by and for the survival of our genes, there would be nothing more than a world of indifferent physical and chemical events that occur inside and outside the brains of living creatures. But that is not the world our mind creates.

The complexities of sexual reproduction have spawned a large array of pleasant and unpleasant feelings that permeate every aspect of our reproductive success: our survival to reproductive age (hunger, pain, etc.), reproduction (beauty, jealousy, etc.), and the care of offspring until they reach reproductive age (love, pride, etc.). As evolved emergent properties of the human brain these adaptive feelings infuse our world with value, adding meaning to mere existence. Viewed from this perspective sexual reproduction is a Faustian bargain foisted upon us by the Red Queen's arms race. It trades potential immortality for a short but meaningful life enriched by the breadth and depth of human sentiments. But would we have it otherwise? After all, in the words of N. Lane, 'A world without sex is a world without the songs of men and women or birds or frogs, without the flamboyant colors of flowers, without gladiatorial contests, poetry, love, or rapture. A world without much interest.'[30]

12 DARWIN'S CARE FOR HUMANITY

Eve-Marie Engels

Introduction

It is often taken for granted that Charles Darwin argued the case for eugenics and Social Darwinism. Both terms were, however, coined after Darwin's death. But did he not advocate eugenic and social Darwinist views before these terms were coined? There are passages in Darwin's work that are usually quoted in order to show that he held such views. I propose a closer and deeper reading of Darwin's text. Darwin has to be read against the background of a variety of contemporary assumptions with which he had to grapple. Above all Darwin's ethical background, his roots in the theory of the moral sense, of sympathy, has to be taken into account. By allowing Darwin to 'speak for himself', I hope to deprive biologistic interpretations of their basis. However, some ambivalences in Darwin's thinking are conducive to misunderstanding and one-sided interpretation.

Darwin's *Origin of Species* and Malthus's Law

Already in his lifetime Darwin's scientific achievement was acknowledged as a *scientific revolution*. Darwin succeeded in explaining the origin of species within the framework of natural science. Species come into being by the *transformation of other species*. Thus he rejects the idea of a special or separate creation of each species by the Creator as well as the idea of the fixity of species. Darwin, however, does not claim to be able to refute by his theory the existence of God. Since this theory has consequences for our general understanding of living nature, including the human being, its impact is not only a scientific but also a philosophical revolution.[1]

Darwin refers to *On the Origin of Species*[2] as 'no doubt the chief work of my life'.[3] The first edition, which came out on 24 November 1859, was sold out almost overnight, and in 1860, a second edition followed. By 1872, Darwin had published six editions of this work. In 1876 the book had been translated into almost every European language.

Darwin names his theory the 'theory of descent with modification through natural selection'.[4] His starting point is the assumption that there is an analogy between the origin of new breeds of plants and animals by domestication (nowadays called 'artificial selection') and that of new species in free nature. In domestication four elements are involved: (1) *individual variation* among organisms of a variety or race; (2) the conscious, intentional *selection* of certain individuals for breeding; (3) the *inheritance* of many of their traits; and (4) the prevention of crossing back by *reproductive isolation*. Breeders select those individuals of a race that have certain traits or characters useful for the breeders' purpose and let them propagate. In the course of many generations the traits, insofar as they can be inherited, gradually prevail or take the form intended by the breeder. In order to maintain these traits it is necessary to avoid the crossing of these individuals with others which lack them. 'The key is man's power of accumulative selection: nature gives successive variations; man adds them up in certain directions useful to him. In this sense he may be said to make for himself useful breeds.'[5]

In free nature there is an analogous mechanism of selection, where the selected traits, however, are useful for an organism itself in a certain environment. Darwin proceeds from the observation that two organisms of the same species are never completely identical. There are always *variations*, however small, and thus also differences in adaptation to an environment. Those organisms of a species whose traits are better adapted to their environment, have a higher chance of survival and thus can more successfully reproduce than the others. This means that a *natural selection* of the better adapted takes place. Those traits which are advantageous for survival can accumulate during generations by inheritance and thus increasingly change, compared to the traits of the aboriginal stock. Thus in the course of long periods of time, from individual variants hereditary varieties, subspecies and finally new species evolve. Darwin advocates a *gradualism* and draws on the *principle of continuity* of natural philosophy.[6] However, this principle is not static any more, it becomes *dynamic* and it stands for a *real relationship* between species.

Thus natural selection not only leads to the dying out of species but also fulfils the *constructive* function of bringing forth new species.

As opposed to artificial selection, however, there is no breeder in nature who purposively chooses organisms for propagation. How can selection be applied to organisms living in a state of nature? Since Darwin did not draw any more on the Creator as the intelligent designer of species, he had to discover some kind of a non-personal natural mechanism fulfilling the function of a *natural* selection.

Here Darwin leaves the analogy between artificial and natural selection. He names this natural mechanism the 'struggle for life' or 'struggle for existence' and draws upon Malthus's law of population. The political economist and clergyman Thomas Robert Malthus points to the disproportion between the arithmetical progression of the means of subsistence (1, 2, 3, 4, 5, 6) and the geometrical progression of the human species (1, 2, 4, 8, 16, 32), when unchecked.[7] Since human populations exhibit by and large stability, however, there must be a mechanism that limits this increase. Malthus assumes the constant effect of *preventive checks* (late marriage or restraint from marriage in connection with sexual abstinence, celibacy) and *positive checks* (common diseases and epidemics, wars, pestilence, plague, convulsions of nature, famine).[8] Malthus terms the preventive checks 'moral restraint', meaning 'a restraint from marriage from prudential motives, with a conduct strictly moral during the period of this restraint'.[9] These two kinds of checks 'are all resolvable into moral restraint, vice, and misery'.[10] For Malthus this principle of population is a natural law imposed by God for the well-being of his creatures.[11]

Reading Malthus's *Essay* on 28 September 1838, Darwin hit on the idea of how to apply the concept of natural selection to free nature.

> Hence, as more individuals are produced than can possibly survive, there must in every case be a struggle for existence, either one individual with another of the same species, or with the individuals of distinct species, or with the physical conditions of life. It is the doctrine of Malthus applied with manifold force to the whole animal and vegetable kingdoms; for in this case there can be no artificial increase of food, and no prudential restraint from marriage.[12]

The scarcity in resources is for Darwin however only one of many causes of this often misunderstood 'struggle for existence'. Even in situations of abundance of food there can be a struggle for existence with differential reproduction. Life may depend on skills like better hiding, faster climbing, smarter behaviour. *Cognitive, social* and *moral faculties* can be as important as *bodily vigour* in this struggle for existence in which only the fittest can survive. From the fifth edition of *Origin of Species* on, Darwin uses the term 'survival of the fittest', coined by Herbert Spencer (1864), in addition to 'natural selection' – 'fit' meaning being adapted to fulfil the demands of a given situation.[13]

For Darwin there is no exception to Malthus's law. And he already includes the human being in his calculation: 'Even slow-breeding man has doubled in twenty-five years, and at this rate, in a few thousand years, there would literally not be standing room for his progeny.'[14]

Particularly the moral or preventive checks described by Malthus play an important role in Darwin's later arguments in his *The Descent of Man* in the context of his care for the future of mankind.

The metaphors 'natural selection' and 'struggle for existence' were subject to much misunderstanding. They can have quite different meanings: *intra*specific as well as *inter*specific competition, and struggle for existence of an individual against environmental dangers (drought, coldness, wetness etc.). Darwin moreover uses the term 'Struggle for Existence in a large and metaphorical sense, including dependence of one being on another, and including (which is more important) not only the life of the individual, but success in leaving progeny.'[15] Nevertheless, the phrase 'struggle for existence' has often been interpreted as a bloody or deadly fight between individuals, races or species. Darwin expresses apprehensions concerning the equivocality of the expression, speculating that the German expression 'Kampf' etc. did not quite express the same notion.[16]

Depending on the situation, the struggle for existence can be coped with by *competition* or *cooperation*. This is particularly outlined later in Darwin's *The Descent of Man*. Mutual aid is a line of Darwinian thought that was pursued particularly in the Russian reception of Darwin.[17] Depending on the situation this struggle manifests itself in many different ways and calls for different ways of coping. Moreover terms like 'struggle for existence' and 'war of nature' were not coined by Darwin but were already current when he published his *Origin of Species*. They were used before, after and independently of Darwin.

Already at the end of the first edition of *Origin of Species* Darwin alludes to the importance of his theory of descent for our understanding of the human being.[18] Although Darwin's conviction of the relationship between the human being and other animals can be traced back to his early *Notebooks* of 1837,[19] he did not publish his *The Descent of Man* before 1871 for fear of adding to the prejudices against his views.

Darwin's *The Descent of Man*: Evolutionary Anthropology and the Evolution of the Moral Sense

Already shortly after the publication of *Origin of Species* in 1859 and in the early 1860s Darwin's theory was applied to account for the appearance of the human being on earth. Many people suspected implications of Darwin's 'Ape Theory' for man. When he finally published his *Descent of Man* in 1871, the world had already been prepared by his supporters Thomas H. Huxley, Carl Vogt, Ludwig Büchner, Friedrich Rolle, Ernst Haeckel and others. It is, however, advisable to read the original Darwin.

Darwin's *Descent of Man* makes him the founder of a comprehensive evolutionary anthropology that includes the physical, mental and moral aspects of the human being. He is open to the multifaceted character of the phenomenon of human morality. Ethics is a special issue in chapter 4, 'Comparison of the Mental Powers of Man and the Lower Animals'.[20] Darwin draws on ethics as a conceptual framework for understanding the human being as a moral being. He focuses on the 'moral sense or conscience' which 'is summed up in that short but imperious word *ought*, so full of high significance'[21] and illuminates it from the perspective of different ethical traditions. These are the English and Scottish philosophers of the moral sense (David Hume, Adam Smith, Alexander Bain),

Kant and his idea of human dignity, virtue ethics as well as historians of ethics (James Mackintosh, William Lecky). He reflects their basic assumption in the light of his *theory of descent* and looks for the evolutionary roots of our moral faculty in the natural history of man, a history that connects us with other animals.

The above described elements of Darwin's theory have consequences for our understanding of man: humans have descended from another, a nonhuman species, in a process of gradual transformation, natural selection being the important mechanism. There exists no 'fundamental difference' between man and his animal progenitors. No matter how large the difference may be, it is certainly 'one of degree and not of kind'.[22]

In *The Descent of Man* Darwin outlines his evolutionary anthropology and his ideas about the evolution of the moral sense.[23] He first adduces bodily similarities (*homologies*, facts from *embryology* and *rudiments*) in order to substantiate his assumption that the human being has evolved from other animals. Then he turns to the cognitive and emotional faculties. The variability of the mental faculties in the individuals of the same species is for Darwin's theory as important as the variability of the bodily structure. In the chapters on the *mental powers of man and animals* he describes a broad range of emotional and cognitive faculties that can be found in humans as well as other animals. Animals of many kinds are social and aid one another in many important ways.[24] These behaviours are rooted in *social instincts*. Because humans have descended from non-human beings that were already invested with social instincts, we enter the world outfit with an evolutionary heritage. An important element of such social instincts is *sympathy*. It 'forms an essential part of the social instinct, and is indeed its foundation-stone'.[25] Darwin explains the emergence of these instincts by natural selection, ascribing to them a function necessary for preserving the community. At the base of the social instincts lie parental and filial affections, gained through natural selection, and the 'feeling of pleasure from society is probably an extension of the parental or filial affections'.[26] Living in close associations is advantageous for survival, for 'individuals which took the greatest pleasure in society would best escape from various dangers; whilst those that cared least for their comrades, and lived solitary, would perish in greater numbers'.[27] Communities with 'the greatest number of the most sympathetic members, would flourish best, and rear the greatest number of offspring'.[28] In all animals, however, sympathy is not directed indiscriminately towards all individuals of the same species, but solely to the members of the same community.[29] Therefore it is probable that also primeval man, as do savages, regard actions 'as good or bad, solely as they obviously affect the welfare of the tribe – not that of the species, nor that of an individual member of the tribe'.[30] At the beginning of man's evolution, humans neither esteemed 'self-regarding virtues' nor did they have the welfare of humanity in general in mind.

The moral sense is 'aboriginally derived from the social instincts, for both relate at first exclusively to the community'.[31] Compared to our early apelike and human progenitors, our instincts are, however, *reduced* in several ways, concerning the quantity, the specialization and the strength of instincts. The condition for the development of genuine morality is this reduction of instincts along with the evolution of reason, judgement and articulate language, the emergence of 'free intelligence' and 'free will'. Nevertheless we still have *instinctive impulses*, which have to be oriented by reason. 'A man who possessed no trace of such instincts would be an unnatural monster'.[32] Darwin by no means reduces man's moral sense to social instincts. Although the 'first foundation or origin of the moral sense lies in social instincts',[33] which are the roots of our moral sense, they alone neither suffice to explain nor to characterize the phenomenon of morality.

> To do good in return for evil, to love your enemy, is a height of morality to which it may be doubted whether the social instincts would, by themselves, have ever led us. It is necessary that these instincts, together with sympathy, should have been highly cultivated and extended by the aid of reason, instruction, and the love or fear of God, before any such golden rule would ever be thought of and obeyed.[34]

Man's moral sense constitutes a qualitatively new capacity. Genuine morality involves consciously made judgements and actions in accordance with ethical principles like Kant's law of morality. This presupposes that an organism's *intellectual faculties* have reached a certain level of development which, according to scientific insights gained so far, man alone possesses. Therefore, according to Darwin, only humans can be moral beings.

Man is for Darwin the 'most dominant animal that has ever appeared on this earth'.[35] Darwin was influenced by Wallace's 'admirable paper' 'The Origin of Human Races and the Antiquity of Man deduced from the theory of "Natural Selection"' of 1864.[36] Two ideas have to be highlighted here. Firstly, Wallace claims that during man's evolution the selection pressure switched from the modification of the body to the improvement of mental, sympathetic and moral faculties.[37] The evolution of articulate language would lead to a still further advancement of the mental faculties. Humans could adapt to the outer world by developing ever new techniques. Man 'would be kept in harmony with the slowly changing universe around him, by an advance in mind, rather than by a change in body', except for the great changes in skull and brain, the 'organ of the mind'.[38] Secondly, a 'grand revolution was effected in nature' without parallel in history by the evolution of humans. The 'true grandeur and dignity of man' is twofold, because 'man has not only escaped "natural selection" himself, but he actually is able to take away some of that power from nature which, before his appearance, she universally exercised'.[39] Man helps the weak and 'saves the sick and wounded from death; and thus the power which leads to the rigid destruction of all animals who cannot in every respect help themselves, is prevented from acting on him'.[40] Wallace's reflections were well suited to Darwin's own ideas about social and moral progress.

Darwin draws on Wallace's ideas but expands them considerably by reconstructing the plausible steps of our social and moral evolution. How could *moral virtues* in contrast to mere social instincts evolve and improve in the course of time? Starting from the social instincts in animals, our apelike progenitors and primeval man, Darwin reconstructs the putative process of their evolution. Groups and tribes, whose members supported each other by mutual aid, had a selective advantage over other tribes and thus could 'supplant' or 'absorb' them, for 'Selfish and contentious people will not cohere, and without coherence nothing can be effected'.[41] This process was repeated many times in history, so the social virtues could spread all over the world. Thus Darwin explains the spreading of social virtues by natural selection of tribes among tribes. The terms 'supplant' or 'absorb' do not necessarily mean that the more successful tribes eliminate the other tribes, for absorption can also mean integration.[42]

This *group-selectionist* approach, however, presupposes the existence of social virtues on the part of the individuals of a community. But how could a large number of members of a tribe *first become* endowed with social and moral qualities, and how could the *standard* of excellence be raised *within* a tribe. For Darwin it seems unlikely that this could happen through natural selection, the survival of the fittest. 'He who was ready to sacrifice his life, as many a savage has been, rather than betray his comrades, would often leave no offspring to inherit his noble nature. The bravest men ... would on an average perish in larger numbers than other men.'[43] Thus, the spreading of moral virtues and the improvement of morals have to take place by other means than natural selection.
Darwin reconstructs some of the 'probable steps' of the evolution of moral qualities from a semi-human condition through that of primeval man and modern savage to civilized humans by mentioning three stages. At the first stage, social action is performed in the hope for getting aid in return by one's fellows when needed, a selfish motive. 'From this low motive he might acquire the habit of aiding his fellows; and the habit of performing benevolent actions certainly strengthens the feeling of sympathy which gives the first impulse to benevolent actions. Habits, moreover, followed during many generations probably tend to be inherited.'[44] The second stage of the evolution of the social virtues is 'the praise and the blame of our fellow-men'.[45] The mutual exchange of praise and blame is primarily rooted in the instinct of sympathy. Setting an example by good deeds for the community and thus becoming a model for imitation was much more effective for reaching larger numbers of people and for improving the standard of morality. The wish for social recognition and glory and the fear of blame were strong incentives. A man 'might thus do far more good to his tribe than by beget-

ting offspring with a tendency to inherit his own high character'.[46] Thus virtues spread out by *cultural transmission*. But morality cannot be reduced to the love of praise and the fear of blame, to accordance with public opinion. Later in evolution, reason, education and religion become increasingly important. The third and highest stage of moral evolution is reached when man has learnt to 'value justly the judgements of his fellows' and might declare 'in the words of Kant, I will not in my own person violate the dignity of humanity'.[47] 'Ultimately man does not accept the praise and blame of his fellows as his sole guide, though few escape his influence, but his habitual convictions, controlled by reason, afford him the safest rule. His conscience then becomes the supreme judge and monitor.'[48] Sentiment, reason and the power of judgement as human abilities as well as education, religion, law and public opinion as external determinants result in the complex entity of the moral sense.

Although the moral sense has its origin in selfish motives, in the course of time an autonomous and true moral motivation for the promotion of the well-being of others and of the community can arise. Darwin rejects ethical egoism and the assumption that 'the foundation of the noblest part of our nature' can be laid in the 'base principle of selfishness'.[49]

In the course of human evolution the quality or standard of morality as well as the number of well-endowed men would increase, and our sympathies would become 'more tender and more widely diffused'.[50] They extend to 'men of all nations and races',[51] as well as 'to the imbecile, maimed, and other useless members of society, and finally to the lower animals, – so would the standard of his morality rise higher and higher'.[52] He considers 'disinterested love for all living creatures' as 'the most noble attribute of man'.[53] When mentioning the gradual extension of morality Darwin refers to Lecky who describes the widening of our benevolence as a 'circle expanding'.[54]

Darwin regrets that the process of regarding others separated from us 'by great differences in appearance or habits ... as our fellow-creatures' takes a long time. 'The very idea of humanity' is for Darwin a 'virtue, one of the noblest with which man is endowed', and

> seems to arise incidentally from our sympathies becoming more tender and more widely diffused, until they are extended to all sentient beings. As soon as this virtue is honoured and practiced by some few men, it spreads through instruction and example to the young, and eventually becomes incorporated in public opinion.[55]

The 'highest possible stage in moral culture' is for Darwin the recognition 'that we ought to control our thoughts'.[56] In his optimistic vision of a far-off future, Darwin assumes that 'virtue will be triumphant'.[57] An essential factor in moral progress is the belief in an 'all-seeing Deity'.[58] Although Darwin abandoned the hypothesis of God for his scientific theory of descent, in his ethics God and religion play an important role. This ethical and social function of the belief in God, however, has to be separated from the question of God's existence.

Moral progress has taken place regarding the subject, the object and thus the standard of morality. On the side of the subject it consists in a progressive evolution from social behaviour based on calculating egoism through conventional morals to autonomous morals based on ethical principles. On the side of the object or receiver of virtuous action it consists in an ever expanding circle from one's own group extending to all members of the same nation, to other nations, all races and finally to animals. We become cosmopolitans. Moral progress, furthermore, consists in regarding not only the welfare, but also the happiness of others. And it consists in regarding not only the community, but also individuals, including 'the imbecile, maimed, and other useless members of society'. In this process our sympathies become more tender and widely diffused and the standard of morality rises higher and higher. Darwin considers moral qualities as 'the highest part of man's nature'.[59] And he dignifies the autonomous value of intellectual and social activities independently of reproductive success. 'Great lawgivers, the founders of beneficent religions, great philosophers and discoverers in science, aid the progress of mankind in a far higher degree by their works than by leaving a numerous progeny.'[60]

Natural Selection and Morals in Civilization

The section '*Natural Selection as affecting civilized nations*' of chapter 5 of *The Descent of Man* explores a variety of subjects that will be crucial in the next decades and centuries. The most important issue is the impairment or even annulation of natural selection in civilization by a variety of different means. These are: (a) morals, (b) modern medicine, (c) certain laws, institutions, customs and traditions, and (d) money.

In this section Darwin writes that most of his remarks are taken from Wallace, Greg and Galton.[61] Darwin does not simply adopt their views, but he presents them and comments on them, expressing his own ideas on these subjects.

These authors evaluate the defeasance of natural selection in different ways. Whereas Wallace describes man's growing power over natural selection by morals as a victory of humanity, Greg is concerned, because he fears that the increase of sympathy and morals has negative effects on the health of the human species and that these results of civilization outweigh its positive aspects. In his article 'On the Failure of "Natural Selection" in the Case of Man' he refers to the 'Darwinian theory of the origin of species'. However he does not quote from Darwin, but from Wallace's article (1864).

Greg considers natural selection as the 'righteous and salutary law which God ordained for the preservation of a worthy and improving humanity'.[62]

Referring to Wallace, Greg points out that the law of natural selection has been 'in many instances almost *reversed*'.[63] It is counteracted by civilization, which 'with its social, moral, and material complications, has introduced a disturbing and conflicting element'.[64] Natural advantages have been replaced by 'artificial and conventional' advantages.[65] Civilization is an upside-down world, in which 'Malthus's prudential check' fails to work in the desired manner. It 'rarely operates upon the lower classes; the poorer they are, usually, the faster do they multiply; certainly the more reckless they are in reference to multiplication'.[66] Here, as in Galton's utopia, Plato may have served as a model. 'The very men whom a philosophic statesman, or a guide of some superior race would select as most qualified and deserving to continue the race, are precisely those who do so in the scantiest measure' whereas the others 'breed *ad libitum*'.[67] Greg describes the double-edged effects of medical science, which is 'mitigating suffering, and achieving some success in its warfare against disease; but at the same time it enables the diseased to live'.[68] The higher average of *life* may result from a lower average of *health*.[69] 'We have kept alive those who, in a more natural and less advanced state, would have died – and who, looking at the physical perfection of the race alone, had better have been left to die.'[70] Greg describes especially three tendencies of the age, that 'run counter to the operation of the wholesome law of "natural selection"': the 'freedom of the individual will', the refusal to 'let the poor, the incapable, or the diseased die' and lastly democracy, which 'means the management and control of social arrangements by the least educated classes'.[71]

According to Galton, Darwin's half-cousin, humans, like any other animal species, can and should be submitted to systematic selective breeding for the purpose of improvement. 'If a twentieth part of the cost and pains were spent in measures for the improvement of the human race that is spent on the improvement of the breed of horses and cattle, what a galaxy of genius might we not create!'[72] Galton was strongly influenced by Darwin's *Origin of Species*, which 'made a marked epoch in my own mental development, as it did in that of human thought generally'.[73]

Yet Galton draws consequences that Darwin himself had not drawn in *Origin of Species*. Galton's writings, mentioned by Darwin in *The Descent of Man*, contain the germ of his later work on eugenics. He draws up a utopia, in which, based on a system of competitive examination, groups of the best ten young females and males are formed 'to embrace every important quality of mind and body' and are encouraged to marriage by a financial and other enticements, like a solemn wedding ceremony in Westminster Abbey, carried out by the sovereign herself.[74] Galton also envisages the distribution of the rising generation into two casts, A and B, A having been selected for natural gifts, B being the 'refuse'. If the marriages in caste A were somehow hastened and those in caste B retarded, in the long run B would be eliminated and replaced by A, 'I believe, we should have agencies amply sufficient to eliminate B in a few generations. I hence conclude that the improvement of the breed of mankind is no insuperable difficulty.'[75] Galton describes practices which are later called 'positive' and 'negative eugenics', positive eugenics being the improvement of the human species by consciously choosing and combining the 'best' individuals, negative eugenics being the restriction or prevention of the propagation of individuals with undesired traits.[76] 1883, one year after Darwin's death, Galton coined the term 'eugenics' for the

> science of improving stock, which is by no means confined to questions of judicious mating, but which, especially in the case of man, takes cognisance of all influences that tend in however remote a degree to give to the more suitable races or strains of blood a better chance of prevailing speedily over the less suitable than they otherwise would have had.[77]

The word 'eugenics' replaces the word 'viriculture', once used by Galton. Galton's ideas about eugenics are also backed up by his criticism of the structure and laws of English society, like the law of primogeniture, 'where riches were more esteemed than personal qualities'.[78] He also criticizes the injustice, which consists in augmented chances of founding a family because of family property. 'The sickly children of a wealthy family have a better chance of living and rearing offspring than the stalwart children of a poor one.'[79]

Eugenics should be 'introduced into the national conscience, like a new religion'.[80]

In his later writings at the beginning of the twentieth century, Galton described the task of eugenics by comparing it with natural selection. The relationship between eugenics and natural selection is one of *cooperation* and *replacement*. Eugenics cooperates with natural selection in terms of their results, the improvement of the human species. But it is also a replacement of natural selection by means that are faster and more humane than natural selection: 'What Nature does blindly, slowly, and ruthlessly, man may do providently, quickly, and kindly.'[81] It falls well within man's 'province to replace Natural Selection by other processes that are more merciful and not less effective'.[82] Galton considers eugenics as a means of preventing the destruction of humans by natural selection that follows from their excessive production. Eugenics rests 'on bringing no more individuals into the world than can be properly cared for, and those only of the best stock'.[83] The first object of eugenics is 'to check the birth-rate of the Unfit ... The second object is the improvement of the race by furthering the productivity of the Fit by early marriages and healthful rearing of their children.'[84]

Darwin picks out ideas of these authors and paraphrases them, but his remarks also mirror his own concern. He compares the breeding of domestic animals with the propagation of humans, 'but excepting in the case of man himself, hardly any one is so ignorant as to allow his worst animals to breed'.[85] On the one hand he fears that the health of the human species is put at stake by civilization; on the other hand he believes in the advantages of civilization for bodily rigour and health. What seems even more important is that he has a clear-cut understanding of moral progress in the sense previously outlined. Our moral sense gradually evolved in a long process of painstaking progressive cultivation and is our most precious good. We could not 'check our sympathy, even at the urging of hard reason, without deterioration in the noblest part of our nature'.[86] Darwin uses the metaphor of the surgeon as a contrasting example for our obligations towards the helpless. We must not behave like the surgeon who 'may harden himself whilst performing an operation, for he knows that he is acting for the good of his patient'. If we, however, 'were intentionally to neglect the weak and helpless, it could only be for a contingent benefit, with an overwhelming present evil'.[87] Therefore we must help the poor and alleviate their misery.[88] According to Sheila Weiss, Darwin is facing a

> personal dilemma in *The Descent of Man*: how can human beings reconcile the inevitable conflict between the humanitarian ideals and practices of the noblest part of our nature with the interest of the race, whose biological efficiency is allegedly impaired by those very ideals and practices?[89]

This is a moral dilemma. Darwin tries to reconcile the preservation and nurture of the 'noblest part of our nature' with the health and welfare of the human species. To prevent the brutalization of our moral sense, we must 'bear the undoubtedly bad effects of the weak surviving and propagating their kind'. Our species is not necessarily endangered by this, for

> there appears to be at least one check in steady action, namely that the weaker and inferior members of society do not marry so freely as the sound; and this check might be indefinitely increased by the weak in body or mind refraining from marriage, though this is more to be hoped for than expected.[90]

Darwin does not call for any laws or restrictions that prevent the weak from marrying and having children. But he hopes that they themselves take responsibility for their renunciation of children. At the end of his book he repeats this once more: 'Both sexes ought to refrain from marriage if they are in any marked degree inferior in body or mind; but such hopes are Utopian and will never be even partially realized until the laws of inheritance are thoroughly known.'[91] Darwin hopes that once breeding and inheritance are better understood, 'we shall not hear ignorant members of our legislature rejecting with scorn a plan for ascertaining whether or not consanguineous marriages are injurious to man'.[92]

During Darwin's lifetime the laws of heredity were not yet known and most people adhered to the doctrine of the inheritance of acquired characters, whose most prominent representative was Jean-Baptiste Lamarck. Virtue and vice, as habits, culturally transmitted traits and pauperism were thought to become inheritable traits after some generations. Darwin also thought that habits could become inheritable traits after some generations. As far as heredity was concerned, Darwinism and Lamarckism were not opposed, and the essential element of Darwin's theory that separated him from Lamarck was natural selection or the survival of the fittest. Neither did Darwin accept Lamarck's progressionism.

Darwin's remark about consanguineous marriages has a concrete background. His concern is not only scientific but also personal. Consanguineous marriages were no rarity in his time. There was a long tradition of first-cousin marriages in Darwin's family, and Darwin himself had married his first cousin Emma Wedgwood. He was concerned that his children might have inherited his diseases. He had ten children, three of them died early. There were no investigations on possible negative effects on the bodily and mental constitution of the offspring in consanguineous marriages. When in 1871 the Census Act was passing through the House of Commons, Sir J. Lubbock and others proposed to have a question inserted concerning the prevalence of cousin marriages. This request was, however, refused by the House of Commons. Darwin's son George thereupon made an investigation on his own, based on available statistics, on circulars, which he sent out and on patient surveys in asylums, initiated by him.[93] The president of the Statistical Society, where the paper was read, 'said that the paper was obviously of that order which had to be very carefully studied and considered before it could be properly discussed'.[94] A presentation and discussion of the investigation and the remarks on its results in the Society's discussion is not possible in the limits of this chapter.

Darwin corresponded with Galton on subjects of eugenics without using this word. Darwin's son Francis thinks that 'from the first he [Galton] had the support of Charles Darwin who never wavered in his admiration of Galton's purpose, though he had doubts about the practicality of reform'.[95] I believe that Francis Darwin underestimates his father's doubts. Darwin was much more careful in his judgements about eugenics than Galton. His reaction to Galton's article 'Hereditary Improvement' (1873) could be read as a more fundamental criticism of Galton's project. Darwin formulates qualms concerning its practicability for practical as well as principled reasons. For him 'the greatest difficulty ... would be in deciding who deserved to be on the register. How few are above mediocrity in health, strength, morals and intellect; and how difficult to judge on these latter heads'.[96] These are questions about the available number of exceptional individuals as well as of the criteria for deciding who gets on the list of selected individuals for the improvement of the human race. Darwin thus touches the problem, that humans widen the range of possibilities for deciding on the future shape of the human being. Whereas natural selection is a blind process, eugenics is a goal-directed action, in which humans themselves decide on the nature of future man in a more specific way than ever before in history. With respect to animals and plants Darwin calls this kind of selection 'methodical selection'.[97] It would ask for quite a new kind of responsibility. Man would have to 'devise for himself a plan',[98] as C. P. Blacker puts it. It would be our self-design, and we would have to decide which traits we want and which we reject. But even if there were enough people above mediocrity and one would wish to promote their precious traits, would these have to be spread by means of directed mating and numerous offspring?

The Descent of Man points in another direction. Darwin's previously mentioned remarks on possible negative effects of civilization for the health of the human species are not meant as an overall criticism of civilization. 'Although civilization thus checks in many ways the action of Natural Selection, it apparently favours the better development of the body, by means of food and the freedom from occasional hardship.'[99] Civilized people are generally physically stronger than savages. Also regarding the intellectual faculties, 'in civilized nations there will be some tendency to an increase both in the number and in the standard of the intellectually able'.[100] The same holds for morality.

'With civilized nations, as far as an advanced standard of morality, and an increased number of fairly good men are concerned, Natural Selection apparently effects but little; though the fundamental social instincts were originally thus gained.'[101] Just the opposite is the case; morality checks the action of natural selection in many ways, as Wallace and Darwin point out. The progress of mankind can be advanced by factors other than successful reproduction. Darwin dignifies the autonomous value of intellectual and social activities independently of reproductive success. 'Great lawgivers, the founders of beneficent religions, great philosophers and discoverers in science, aid the progress of mankind in a far higher degree by their works than by leaving a numerous progeny.'[102] Patrick Tort calls this the 'reversive effect of evolution'.[103] Natural selection has been set out of force to a high degree.

In his correspondence with Galton, Darwin admits that 'the object seems a grand one; and you have pointed out the sole feasible, yet I fear utopian, plan of procedure in improving the human race'. But can a utopian plan be feasible? This remark was followed by 'one or two minor criticisms', which were not so minor.

There are nevertheless passages in Darwin's text where he seems to plead for the necessity of natural selection in civilization. One of them is at the end of his book.

> The advancement of the welfare of mankind is a most intricate problem: all ought to refrain from marriage who cannot avoid abject poverty for their children; for poverty is not only a great evil, but tends to its own increase by leading to recklessness in marriage. On the other hand, as Mr. Galton has remarked, if the prudent avoid marriage, whilst the reckless marry, the inferior members tend to supplant the better members

> of society. Man, like every other animal, has no doubt advanced to his present high condition through a struggle for existence consequent on his rapid multiplication; and if he is to advance still higher, it is to be feared that he must remain subject to a severe struggle. Otherwise he would sink into indolence, and the more gifted men would not be more successful in the battle of life than the less gifted. Hence our natural rate of increase, though leading to many and obvious evils, must not be greatly diminished by any means. There should be open competition for all men; and the most able should not be prevented by laws or customs from succeeding best and rearing largest number of offspring.[104]

This passage played a role in Darwin's correspondence with defenders of artificial birth control, a growing topic in the nineteenth century.[105] Darwin corresponded on this issue with Charles Bradlaugh in the context of the Bradlaugh–Besant trial (1877–8) and with George Arthur Gaskell (1878).

Bradlaugh was a British Freethinker, Neo-Malthusian and the founder of the Malthusian League, the first organization to advocate birth control as solving the problems of the poor. In England and other countries the population increased rapidly and caused an overcrowding of the poor in town and country. An 'excessive number of births was not only a financial burden on young couples but also a strain on the health of young mothers'.[106] It was also often a death-sentence for the children. Over-large families had to put their children too early to work to make their living. Also 'baby-farming' was popular in England. Three-year-old children were employed as 'gaffer or ganger' over groups of eight younger babies. They were in charge of putting the bottle in their mouth as soon as they woke up. Baby-farming was 'simply a veiled form of infanticide'.[107] Malthus had proposed preventive checks to avoid the misery and evil ensuing from such positive checks. Late marriage, combined with moral restraint, sexual abstinence, should substitute some of the above mentioned positive checks. Artificial contraception was condemned by him:

> Indeed I should always particularly reprobate any artificial and unnatural modes of checking population, both on account of their immorality and their tendency to remove a necessary stimulus to industry. If it were possible for each married couple to limit by a wish the number of their children, there is certainly reason to fear that the indolence of the human race would be very greatly increased; and that neither the population of individual countries, nor of the whole earth, would ever reach its natural and proper extent.[108]

Neo-Malthusians however rejected this kind of preventive checks because one 'would only replace one set of evils by another'.[109] Late marriage with sexual abstinence would promote prostitution. 'Celibacy is not natural to men or to women; all bodily needs require their legitimate satisfaction, and celibacy is a disregard of natural law.'[110] Another serious difficulty was the one, also mentioned by Greg and Galton, that 'the best of the people ... would remain celibate and barren, while the careless thoughtless, thriftless ones would marry and produce large families'. The 'more thoughtful ... have to pay heavy poor-rates for the support of the thoughtless and their families'.[111] Annie Besant concludes, that 'the preventive check proposed by Malthus must therefore be rejected, and a wiser solution of the problem must be sought'.[112] The work of the utilitarian philosopher John Stuart Mill played an important role for the defence of artificial birth control in these debates.

Whereas Malthus's Law, the tendency of living beings to increase faster than the means of subsistence, was accepted by Neo-Malthusians as an adequate description of the situation, they rejected the adequacy of positive and preventive checks as means of population regulation. 'Ought we to leave the sickly to die? Ought we to permit infants to perish unaided? Ought we to refuse help to the starving? These checks may be "natural", but they are not human; they may be "providential", but they are not rational.'[113] Positive and preventive checks should be substituted by 'scientific checks'.[114] Artificial contraception was the solution based on science and reason, and the poor had to be informed about it in a legal way by books which were publicly available.

Fruits of Philosophy: The Private Companion of Young Married People, by a Physician, written by the American doctor and founder of American contraception medicine, Charles Knowlton, and published anonymously in 1832 in New York, was such a book. From 1833 to 1877, 40,000 copies were sold by different publishers in England before the book was banned and the publisher Charles Watts was prosecuted. Bradlaugh and Besant published the book with a new preface in their own Freethought Publishing Company in London and were thereupon arrested and tried; however, they won the case.[115] The trial marked the beginning of a public discussion on contraception and family planning and the availability of birth-control knowledge to all, 'not just to the wealthy few'.[116]

When arrested, Bradlaugh wrote Darwin a letter asking him for help, hoping for Darwin's support before court. Darwin, however, refused his support. If he showed up at court, he 'should be forced to express in court a very decided opinion in opposition' to Bradlaugh and Besant. In his letter he points to the above quoted passage on the advancement of the welfare of mankind[117] and writes:

> When the words 'any means' were written I was thinking of artificial means of preventing conception. But besides the evil here alluded to I believe that any such practices would in time spread to unmarried women & w[d] destroy chastity on which the family bond depends; & the weakening of this bond would be the greatest of all possible evils to mankind; & this conclusion I sh[d] likewise think it my duty to state in Court; so that my judgment, would be in the strongest opposition to yours.[118]

Darwin's letter is particularly striking because the reasons for Darwin's claim that 'our natural rate of increase, though leading to many and obvious evils, must not be greatly diminished by any means', which he gives in his book and in his letter, shed light on the range of Darwin's different concerns. Reading the passage from his publication we get the impression that Darwin's main interest is an overall improvement of mankind by the successful reproduction of more gifted individuals through competition and natural selection. He does not mention contraception as an example of the means that should be avoided, but certain laws and customs, which prevent, that the most gifted individuals succeed best and rear the largest number of offspring. As examples Darwin points to primogenitures with entailed estates – although some 'compensatory checks intervene' here – and the military policies. The 'finest young men' have to serve in the military and are prevented from marrying 'during the prime of life' whereas 'the shorter and feebler men, with poor constitutions, are left at home, and consequently have a much better chance of marrying and propagating their kind.'[119] Regarding the second example Darwin refers to an article by Heinrich Fick, a professor of law, with whom he also had a correspondence about trade unions. Darwin pleads for competition among workers, because the exclusion of competition seemed to him 'a great evil for the future progress of mankind'.[120] But also here he was optimistic, because 'Nevertheless under any system, temperate and frugal workmen will have an advantage and leave more offspring than the drunken and reckless.'[121] These examples show that Darwin criticizes laws by which a certain social group is privileged independently of the achievements of its individuals. He pleads for equal opportunities for rich and poor, for justice. Competence should count, no matter what the social and familial background is. Darwin has the welfare of the human species in view. Under the existent laws, property and health are not necessarily connected with each other.

Besant quotes Darwin's above mentioned passage – interestingly she replaces Darwin's phrase 'advancement of the welfare of mankind' by 'enhancement of the welfare of mankind' – objecting that 'Mr. Darwin forgets that men have qualities which the brutes have not, such as compassion, justice, respect for the rights of others – and all these, man's highest virtues, are absolutely incompatible with the brutal struggle for existence.' 'Scientific checks to population', that is contraception, 'would just do for man what the struggle for existence does for the brutes: they enable man to control the production of new human beings'. Without using the word, Besant points to the potential *eugenic* effects of these 'scientific checks to population': 'those who suffer from hereditary diseases ... might marry, if they so wished, but would preserve the race from the deterioration which results from propagating disease'.[122]

It is not correct to accuse Darwin of forgetting that man is a being capable of morality. As outlined in the previous section, we could not 'check our sympathy, even at the urging of hard reason, without deterioration in the noblest part of our nature'.[123] Therefore we have to care for the weak and sick who otherwise will become the victims of the struggle for life.

If we call this Social Darwinism, it is at least not the rude and cruel Social Darwinism usually identified with this phrase. Darwin advocated and practised private charity, and it would call for more research to spot his final attitude towards poor laws. His mentioning of these laws in *The Descent of Man* is not significant enough, because he refers to them in the context of his presentation of Greg's and others' views. Also a more thorough analysis would be necessary for presenting in detail the pros and cons of the poor laws and trade unions for the people concerned, the poor.[124]

In his letter to Bradlaugh, Darwin bases his rejection of artificial contraception not on the fear that by the reduction of the birth rate the struggle for life would be checked and thus could not exercise its power of improving mankind. Although artificial contraception could also be used as a means of avoiding the deterioration of the human species, Darwin rejects it. In fact, he fears the destruction of chastity as the basis of the family bond by the introduction of artificial birth control. The weakening of the family bond would be the 'greatest of all possible evils to mankind'. Darwin is concerned about the loss of morality and of the social and institutional security provided by the family. In his letter he highlights moral and social concerns about the future of mankind and the threat of moral and social values. That this is his major concern is also supported by the fact that goes on by surpassing the value of natural selection, emphasizing that the highest part of man's nature are his moral qualities.

> Important as the struggle for existence has been and even still is, yet as far as the highest part of man's nature is concerned there are other agencies more important. For the moral qualities are advanced, either directly or indirectly, much more through the effects of habit, the reasoning powers, instruction, religion, etc., than through Natural Selection; though to this latter agency may be safely attributed the social instincts, which afforded the basis for the development of the moral sense'.[125]

Although Darwin did not share Malthus' physico-theological background any more, in one way he proves to be a good Malthusian. Malthus feared immorality and the indolence of the human race by the introduction of artificial contraception.

Moral concerns are also the tenor of Darwin's correspondence with George Arthur Gaskell, which was printed in Jane Clapperton's *Scientific Meliorism and the Evolution of Happiness*. Gaskell wrote to Darwin objecting to his following argument from *The Descent of Man*:

> Natural selection follows from the struggle for existence; and this from a rapid rate of increase. It is impossible not to regret bitterly, but whether wisely is another question, the rate at which man tends to increase; for this leads in barbarous tribes to infanti-

> cide and many other evils, and in civilized nations to abject poverty, celibacy, and to the late marriages of the prudent. But as man suffers from the same physical evils as the lower animals, he has no right to expect an immunity from the evils consequent on the struggle for existence. Had he not been subjected during primeval times to Natural Selection, assuredly he would never have attained to his present rank.[126]

In a letter to Darwin of 13 November 1878, Gaskell objects to Darwin's statement, that man 'has no right to expect an immunity from the evils consequent on the struggle for existence'. Gaskell thinks

> from the advance of civilization, which is so much a conquest over nature, and the growth of altruism, we have reason to hope for this immunity; and as I now think we can have it without any deterioration of race and decline of virtue, I am free to think it wise to regret the continuance of the pressure of population on comfort and subsistence.[127]

Gaskell points out that 'two important laws of Race' are now in action in addition to Darwin's law of natural selection, which destroy the action law of natural selection, 'and the last, which is now in the first stages of evolution, annuls as it grows the action of the two preceding ones'. These laws are 'in their natural order of sequence in evolution' 'First, the 'Organological Law–Natural Selection, or the Survival of the Fittest. Second, the Sociological Law–Sympathetic Selection, or Indiscriminate Survival. Third, the Moral Law–Social Selection, or the Birth of the Fittest'.[128] Gaskell calls the law of natural selection as 'a law of *destruction* and survival', whereas the law of sympathetic selection is 'a law of *protection* and survival'.[129] Gaskell formulates the second law from a consideration especially of chapters 3 to 5 of Darwin's *The Descent of Man* as well as of the writings of Spencer, Wallace, Galton, Greg and others: species continued to flourish although natural selection had been defeated, and the power was sympathy. The third law, the 'Birth of the Fittest' is based on the possibility of artificial birth control. Gaskell firmly believes that this third law 'is destined to act a most beneficent part in the future of mankind'.[130] His view of the evolution of the third law is based on

> the growing opinion that it is wrong for consumptive people and persons inclined to insanity and epilepsy to marry; the opinion, becoming more and more prevalent, that it is wrong to have more children than can be brought up well; the opinion that celibacy is an evil, and that asceticism is absurd; that the sexual passion is at the spring of much that is noble in life, and is nothing to be shamed of, but requires only to be regulated ... the conclusion that procreation is perhaps of all social actions the most important, and ought therefore to be most seriously regarded, and effected only under moral conditions; the opinion that tendency to vice is hereditary, ... And finally I may refer to the present painful conflict between reason and sympathy relative to the preservation of the weak and incompetent while they propagate their stock to the injury of posterity.[131]

According to Gaskell, 'the birth of the fittest offers a much milder solution of the population difficulty, than the survival of the fittest and the destruction of the weak'.[132] The suffering of the sick and weak is prevented because their birth, their existence, is prevented in the first place.

Darwin answers that the second law, 'sympathetic selection, or indiscriminate survival' appears to be largely acted on in all civilized countries and thinks 'that the evils which would follow by checking benevolence and sympathy in not fostering the weak and diseased would be greater than by allowing them to survive and then to procreate'.[133] For Darwin, Gaskell touches on a much larger and more important issue than reproduction. It is a question of preserving and fostering our moral sense by caring for the weak and diseased, even if we thus promote their procreation. Our moral sense is a precious achievement of human evolution which we must not endanger.

Concerning the third law, artificial checks to population growth, Darwin 'cannot but doubt greatly, whether such would be advantageous to the world at large at present, however it may be in the distant future'.[134] He mentions two reasons for rejecting artificial birth control. The first one is a general political one: if artificial checks to population growth had existed in Britain during the last centuries or for a shorter time, the world would be in a very different state, 'when we consider America, Australia, New Zealand, and South Africa! No words can exaggerate the importance, in my opinion, of our colonization for the future history of the world.'[135] In his *Journal of Researches* he describes the achievements of missionaries in Tahiti, who effected the abolition of human sacrifices, infanticide, bloody wars, and the reduction of dishonesty, intemperance and licentiousness by the introduction of Christianity. Darwin is convinced of the superiority of British culture and civilization. The second reason is his concern about the possible deterioration of female and general morals, as already expressed in his letter to Bradlaugh. 'If it were universally known that the birth of children could be prevented, and this was not thought immoral by married persons, would there not be great danger of extreme profligacy amongst unmarried women, and might we not become like to [*sic*] "arreois" societies in the Pacific?' The arreois societies were promiscuous societies with a high rate of infanticide. Gaskell objects that these societies are not representative, because due to their 'libertine and selfish natures' they are 'social suicides'.

Darwin obviously rejects artificial checks on reproduction, a biotechnological solution of birth control, like that propagated by the Neo-Malthusians. Where birth control is appropriate, he pleads for preventive checks like those, proposed by Malthus, for chastity and late marriage. And he calls for refraining from marriage if there is a risk for the descendants of inheriting their parents' diseases. The sick and poor must not be abandoned to their fate, to the misery of positive checks. In order to avoid the deterioration of our moral sense, the noblest part of our nature, Darwin pleads for alleviating the effects, which the struggle for life has on the weak and sick members of society, by caring for them. For Darwin benevolence and sympathy play an important role in our treatment of needy people.

Darwin was a humanist. He did not only express humanitarian ideas in his work, but he was a convinced abolitionist and considered slavery as 'great crime'.[136] This attitude was a family tradition. Both his grandfathers, Erasmus Darwin and Josiah Wedgwood I, publicly protested against slavery, and Charles Darwin also detested it, as is expressed in his letters and his journal of researches. As opposed to many contemporaries, he was convinced of the unity of the human species. The 'so-called races of man' are not different human species for him, but they belong to just one human species. Darwin was a monogenist, he rejected polygenism and thought that races had evolved after the evolution of the human species, after the 'rank of manhood' had been reached.[137]

Darwin's phrases 'higher' and 'lower' races are not meant as biological categories but as the description of the cultural and moral level of certain savage tribes. Darwin rejects infanticide and considers certain ways of treating old people as brutish. He made the acquaintance of savages during his *Beagle* voyage and by reading the works of Lubbock, Captain Cook and others. Darwin emphasizes the compliance and plasticity of the three young Fuegans, who had lived for some time in England and were with him on the *Beagle* on their way home to Tierra del Fuego; and he experienced the broad range of Fuegans's plasticity when he could compare them with their savage fellow-men.

Together with his wife, Emma, Darwin was also active in animal protection. For Darwin, the moral qualities are the 'highest part of man's nature', and natural selection decreases in human history in its importance as a motor of progress. 'With highly civilized nations continued progress depends in a subordinate degree on Natural Selection; for such nations do not supplant and exterminate one another as do savage tribes.'[138]

Conclusion

Darwin's position in his argument with Neo-Malthusians is striking because it shows that he firmly keeps up with traditional values. His pleads for late marriage or refraining from marriage; for caring for the weak members of society; his attitude towards artificial contraception; to a biotechnological solution of birth control; as well as his high estimation of morality, expressed in *The Descent of Man*, manifest that he not only cares for the health of the humans species but also for the maintenance of a certain moral value system. For Galton, the effect of Darwin's *Origin of Species* 'was to demolish a multitude of dogmatic barriers by a single stroke, and to arouse a spirit of rebellion against all ancient authorities whose positive and unauthenticated statements were contradicted by modern science'.[139] Within certain limits this may be true for the effects of Darwin's general theory of descent. Regarding Darwin's own intentions with respect to ethics, however, a close reading of *The Descent of Man*, a consideration of his correspondence and of his personal commitments to his family and community convey a different impression. Darwin defended a humanitarian ethics and the maintenance of the family. Others like Alexander Tille and Wilhelm Schallmayer have criticized Darwin for adhering to humanitarian ideals and applied his theory – or what they took to be his theory – in a much more radical way than Darwin himself. He was aware of the ethical achievements and advantages of culture, which were precious goods that must not be endangered.

Acknowledgements

I cordially thank my student assistants Michael Botsch, Johanna Edel and Katharina Meyer-Borchert for their support in providing literature. I also thank Johanna Edel and Michael Botsch for their assistance in preparing the index.

Eugenics was the guardian of the continuation of the human species as a Western civilization. Engendering the survival of the fittest by policy required restructuring the society and a biological benchmark that turned on a binary categorization premised on reproductive fitness versus reproductive unfitness. Eugenics placed reproduction under collective aegis eliding individuality. In the words of the Oxford philosopher, and mainline eugenicist, F. C. S. Schiller, eugenic reproduction hinged on 'national self-selection'.[9] According to Galton, 'the term 'individuality' is in fact a misleading word'.[10] Dr Slaughter, a leading eugenic propagandist, made it clear that 'the only obstacle to "the elimination of the unfit" is the influence exerted by other catch words which mean nothing at all ... such as "humanity" and "the sacredness of the individual in a democratic society"'.[11]

Eugenic reproduction entailed direct correlation to groups of individuals encouraged to breed to benefit the community, and individuals whose reproduction was regarded undesirable by virtue of carrying degenerative traits. Galton singled out eugenic qualities of primary importance as the ones he associated with his social class, 'special faculties as broadly distinguish philosophers, artists, financiers, soldiers, and other representative classes ... sound mind and body, enlightened, I should add with an intelligence above the average'.[12] Unfitness covered a wide spectrum of mental and social traits, supposedly hereditary, ranging from pauperism to mental illness: 'craving for drink, or gambling, strong sexual passion, a proclivity to pauperism, to crimes of violence, and to crimes of fraud'.[13]

In line with the tradition ingrained in popular culture of regarding crime and mental disorder as one and the same, the 'category problem' in eugenics lumped together the pathological and pauperism inflected by notions of vice and degeneration. As Elazar Barkan and Simon Schaffer have documented, class bias at the core of British eugenics spilled into an obsession with race differentiation and mental health that persisted well into the mid- twentieth century.[14]

The roots of the concept of 'hereditary taint' can be traced to early nineteenth-century medical and layman discussions on procreation in Britain and the United States, in particular to phrenologists like Orson Fowler. As John Waller has outlined, the inverse correlation between degenerative traits, ranging from venereal disease to alcoholism, and national-racial health originated in this context.[15] This way of thinking about the 'category problem' centring on feeblemindedness coalesced in the early twentieth century with a strand stemming from Henry Goddard's work. His ideas, inspired by Binet's intelligence tests conjoined with Mendelian speculations, crossed from the United States into Britain to strengthen the assumption that the condition of idiocy or the 'moron' was a form of undeveloped humanity.

Eugenics: Scientific Beehive

Galton had aspired to 'place eugenic thought, where possible, on a strictly scientific basis'.[16] Eugenics sought legitimacy in natural science and social research. By virtue of taking heredity – whose mechanisms and units of transmission were yet to be discovered – as its central concern, it contributed to the development of a body of knowledge that became integrated into modern science. Eugenics was vested with scientific respectability and became an interdisciplinary locus with the establishment in 1911 at University College London of the Department of Applied Statistics, harbouring the Galton Laboratory for the Study of National Eugenics, merged with Pearson's biometric laboratory, and the Galton Professorship of Eugenics.

A combination of factors such as professional self-interest and the need for eugenicists to develop a heuristic toolbox in support of their beliefs, coupled with socio-political circumstances explain the deep inroads of eugenics in the world of science. The Department of Applied Statistics trained researchers from England, the Continent, the United States, India and Japan, and a stream of Cambridge and Oxford graduates, working in various fields: mathematics, medicine, biology, anthropometry, criminology, psychology, economics and agriculture.

On the other hand, the transformation of eugenics into a topic of public interest relevant to the learned society, reformers, and political leaders was achieved through the creation of the Eugenics Education Society (EES) in 1907. The society changed aims regularly, and although its membership never achieved large numbers, it became a powerful propagandistic forum. Its leaders possessed a myriad of contacts in academia and the London politico-medical establishment and engaged in social activism and legislative pressure through the publication of pamphlets, sponsorship of research and lobbying of governments.[17]

Despite a significant overlap in affiliation between the EES and people from the Galton Laboratory, Pearson as well as many other eugenicists never joined the Eugenics Society. Lindsay Farrell showed that scientists and university lecturers figured prominently in the British Eugenic Movement; two thirds of these were biological or social scientists.[18]

Alexander Carr-Saunders, secretary of the EES, was one of the leading scientists trained by Pearson in statistics for his work on overpopulation. The main methodological framework of the EES was the pedigree structured by Carr-Saunders, Greenwood, Lidbetter and Tredgold 'as a network of relationships demonstrating inheritance of defect in terms of the biological connections within a social class'.[19] The statistical correlations covered the following categories: insane pauper, feeble-minded or idiot pauper, tuberculous, epileptic, blind, still-born, chronic pauper, occasional pauper, pauper child, medical relief, physically unsound, illegitimate, born in Workhouse, illegitimate and born in Workhouse, died in infancy.[20]

When research rendered the pedigree vulnerable, Ronald Fisher tried to devise a statistical variable accounting for heredity and environment as causes of pauperism, an enterprise that proved unsuccessful. While in Cambridge, Fisher founded the University of Cambridge Eugenics Society, an audience attentive to his complaints that society was not stratified eugenically so that 'we are breeding more from the worse than from the better stocks ... socially lower classes have a birth rate ... in excess of those who are, on the wholly, distinctively their eugenic superiors'.[21]

Fisher's contribution to evolutionary theory sealed the episode that drew attention to statistical models in the study of population and heredity, and to eugenic considerations: the controversy over the genetics of evolutionary change. The biometricians focused on descriptions of phenotypic resemblance premised on inheritance through continuous blending and variation. The opposing Mendelian geneticists were concerned with the elusive underlying genotype conceived of as a theoretical entity, an 'invisible', which passed through random distribution from parent to offspring.[22]

William Bateson, who championed Mendelism for Cambridge and coined the term genetics, admonished against genetic engineering for purposes of racial enhancement. Eugenics was driven by perennial ambitions, therefore it would 'be very surprising indeed if some nation does not make trial of this new power. They may make awful mistakes, but I think they will try'.[23] His 1919 Galton lecture was a call for commonsense in racial problems. Forewarning of alliances between pure and applied science, Bateson emphasized that 'Genetics are not primarily concerned with the betterment of the human race or other applications, and I am a little afraid that the distinctness of our aims may be obscured.'[24]

Yet, Bateson's faith in Goddard's theory cemented his conviction that in the case of 'parents, both of gravely defective or feeble mind, in the usual acceptance of that term ... no one can doubt that the right and most humane policy is to restrain them from breeding, and I suppose the principle of the Act before Parliament for the institution of such a policy will have general approval'.[25]

As the campaign for eugenic legislation gained steam, his successor Reginal Punnett, an ardent eugenicist admirer of Goddard, sought better insight into the theory of feeblemindedness, based on a simple recessive Mendelian trait, which he suspected was flawed. Aware that most deleterious genes are hidden in seemingly normal carriers, so that selection has little effect in trying to efface them from the gene pool, Punnett asked G. H. Hardy to calculate how long it would take to eliminate the condition of feeblemindedness. Hardy estimated it would take near a millennium. Sterilization could not wipe out the feebleminded, therefore 'some method other than that of the elimination of the feebleminded themselves must eventually be found'.[26]

However, the spin-off from segregation to sterilization chafed eugenicists like Haldane. The eugenics of Haldane was given a utopian vision in *Daedalus or Science and the Future*, the topic of his 1923 address before the Cambridge Heretics expanded it in a booklet predicting an increased role of science in moulding human affairs. The disengagement of sexual love from reproduction conjoined with the explicit scientific statement of 'ectogenesis', the culture of human embryos in the laboratory, earned the first genetic engineer his flying colours. A signer of the *Geneticists' Manifesto*, Haldane contributed to the reform of eugenics aware that 'students of heredity' were aloof from the movement, which was undermined by the propagandists. The progress of biology, Haldane wrote, will force us to accept that men are innately unequal, and the scientific state would make its first business to investigate this inequality.[27] But eugenics as a practical programme was both scientifically and socially premature.

The most consequential critique of mainline eugenics was pioneered in Lionel Penrose's cunning demonstration that the cluster of epileptics, inebriates, recidivists, deviants and prostitutes expressing hereditary attributes of the lower orders involved circular reasoning. Eugenicists had little contact with the realities of the world where mental defectives were confined. Supported by the EES Darwin Trust, Penrose engaged in the first serious empirical study of the causal roles played by heredity and other possible factors. Ignorance was singled out as the reason why mental disease was assigned more frequently to heredity than physical disorders.[28] Upon taking the Galton chair, Penrose jettisoned eugenics, a term that he replaced by 'human genetics'.

The affinity between eugenics and biology tends to conceal the inroads made by eugenics in a wide spectrum of scientific disciplines. A number of prominent social scientists were diligent EES members: 'the psychologists Cyril Burt, William McDougall and Charles Spearman; the sociologist Patrick Geddes; the political scientists Lowes Dickinson and Harold Laski; the anthropologists A. C. Haddon and C. G. Seligman; and the economist John Maynard Keynes'.[29] A radical eugenicist and a pupil of Pearson, Laski was categorical that 'the different rates of fertility in the sound and pathological stocks point to a future swamping of the better by the worse. As a nation, we are faced with race suicide.'[30]

McDougall's research in psychology was motivated by his eugenic commitment to secure 'improvement of mental sanity and vigour and the level of intellectual and moral efficiency in the human stock ... and especially it will require to know as exactly as possible the mental endowments of the progeny produced by the crossing of these subraces'.[31] Cyril Burt's work on psychometrics, inspired by McDougall and Pearson, was instrumental in substantiating hereditary causation for the eugenic cause which he proclaimed in nationalistic terms: 'Mental inheritance, then, not only moulds the character of individuals; it rules the destiny of nations.'[32]

The backbone of eugenics overlapping heredity, class, and intelligence was especially reliant on another pupil of Pearson, Charles Spearman. Spearman saw in psychological tests an instrument to boost the scientific credibility of eugenic hereditarianism and stressed that whereas 'the development of scientific abilities is in large measure dependent upon environmental factors, that of general ability is almost wholly governed by heredity'.[33]

The frontispiece first edition of the *Annals of Eugenics* (1925) displayed a picture of Malthus as the iconic signifier of the flow of ideas between biology and economics and by extension eugenics through the problem of population. Economic considerations pervaded the eugenic discourse which in turn lured a number of economists. Keynes, like Marshall, was an unreserved believer in hereditarianism, but such convictions are not reason enough to explain his eugenic commitment and his positions as the treasurer of the University of Cambridge Eugenics Society and an officer of the EES council. Nonetheless, personal ties that bound Keynes to the intellectuals and socialites that embraced eugenics, and his reform ideals together with his interest in population, go a long way to explain his involvement in the movement. Keynes held Mathus in great admiration and drew heavily on population to substantiate the arguments set forth in *The Economic Consequences of the Peace* (1920), the topic of the Galton lecture he delivered in 1937. The lecture expressed concerns that declining population might trigger inadequate demand for capital and under consumption, hence compromising standards of living.[34] As John Toye outlined, Keynes was very interested in birth control, opposed pronatalist eugenic policies and endorsed the eugenic attitude on immigration laws as 'protective measures against injury at the hands of more prolific races'.[35]

In the full-fledged eugenic discourse of Arthur Pigou socio-economic improvement was bound both to sanitarian proposals and biological purges. Despite his suspicion of hereditarianism, Pigou held a strong conviction that certain classes of unfit pauper should be prevented from breeding, hence 'permanent segregation' or 'sterilization' was essential to shield society from 'contagion'.[36] To curtail the social costs of parentage, the 'obviously unfit, those afflicted with definitive hereditary taints, the imbeciles, the idiotic, the sufferers from syphilis and tuberculosis, should be authoritatively restrained'.[37] As 'sterilisation can be affected in either sex by a simple operation ... Has not the time come when, with due safeguards and under proper restrictions, this method of social improvement could be recognised and employed?'[38]

Genius and Degeneracy: The View from Intelligentsia

The world eugenics sought to amend was evocatively depicted by Virginia Woolf: 'On the towpath we met & had to pass a long line of imbeciles. The first was a very tall young man ... one realised that every one in that long line was a miserable ineffective shuffling idiotic creature ... it was perfectly horrible. They should certainly be killed.'[39] As Angelique Richardson remarked, eugenics 'found their most virulent expression not in legislation or public policy, but in popular and intellectual discourses'.[40] Donald Childs further noted that 'the eugenics of some writers was notorious even in their own day'.[41]

The melding of eugenic ideas, feminism, pacifism and sexual liberation in the lives and views of the individuals that gravitated around Bloomsbury express the contradictions of eugenics in the realm of the arts, literature and reform. H. G. Wells, who turned out to hear Galton's lecture before the Sociological Society in 1904 with Bernard Shaw, aired his desolation at the appalling picture 'of this vicious helpless and pauper masses' where the railway had spread, in Chicago and New York as vividly as in London.[42] Given their characteristic weaknesses detrimental to the civilizing fabric, to 'give them equality is to sink to their level, to protect and cherish them is to be swamped in their fecundity'.[43]

As G. R. Searle has outlined, Shaw's obsession with further racial improvement culminated in the 'address to the central branch of the Eugenics Society in 1910, where his tongue-in-cheek commendation of the lethal chamber and the methods of the stud farm caused a memorable scandal'.[44] Such views, as Childs suggests, were consistent with ideas expressed in fiction that 'extermination must be put on a scientific basis if it is ever to be carried out humanely and apologetically as well as thoroughly'.[45]

Bertrand Russell strayed from his role of guardian of individual liberties in *Marriage and Morals* where he emphasized the eugenic ideal of improving the biological character of mankind. Birth control was hitherto the more practicable and humanitarian policy but not a solution for the problem of the feeble-minded, for such 'women, as every one knows, are apt to have enormous numbers of illegitimate children, all, as a rule, wholly worthless to the community ... would themselves be happier if they were sterilized ... same thing, of course, applies to feeble-minded men'. [46] Russell admitted that despite 'dangers in the system, since the authorities may easily come to consider any unusual opinion or any opposition to themselves as a mark of feeble-mindedness. These dangers, however, are probably worth incurring ... the number of idiots, imbeciles, and feeble-minded could, by such measures, be enormously diminished'.[47] As Stephen Heathorn has remarked, Russell's eugenic discourse, which caused bewilderment and brought around several conjectures, was ubiquitous in his inner-circle.[48]

The traits of hereditary genius sketched by Galton had been drawn from a survey that included Woolf's acquaintances, namely the Potter sisters. One should have thought that Woolf was likewise endowed with fitness for reproduction despite her fragile physical and mental health. In the view of Sir George Savage, the Royal Society of Medicine delegate to the first International Congress of Eugenics, neurosis in women issued from the better strains could be bred out through reproduction. Donald Childs suggests that in all likelihood Savage regarded Woolf's mental illness as a by-product of genius given that he advised her to have children. The line between hereditary taint and hereditary genius was a blurry one after all. But Woolf's aspirations to motherhood, however, were dashed by her husband's eugenic arguments, an excuse to justify for his dislike of children.[49]

Reproduction and Legislation: Enacting Eugenics

In Britain, the eugenic movement never succeeded in appropriating institutional structures to enact large-scale eugenics as it happened in other countries. However, the arguments deployed in the campaign for voluntary sterilization are of interest to understand the forces driving the identification and segregation of individuals or families tagged unfit for reproduction.

Eugenics was bound to a meritocracy with a foothold in both left and right wing politics. The way they perceived the world was expressive of a particular social experience captured in the way Beatrice Potter Webb, a muse and admirer of Galton, categorized her circle: either as As – Artist, Anarchist and Aristocrat; or Bs – Benevolent, Bourgeois and Bureaucratic. Keynes, a close friend of Beatrice might well be a B. Bertie Russell was an A. The inner-circle of Beatrice included Haldane and Laski, as well as a number of Bloomsbury people; she befriended the Churchills, and was invited by Lloyd George to sit on the Reconstruction Committee.[50]

An aspect of the argument for negative eugenic enactment in Britain turned on the burden the destitute put on society by resorting scurrilously to relief structures: 'the feeble-minded woman, or the woman who is mentally and morally degenerate without being actually imbecile ... treat the local workhouse or Poor Law infirmary simply as a free maternity ... as a nation we are breeding from our inferior stocks'.[51] Despite agreement as to who constituted the category problem, Sidney Webb, who favoured institutional reform, dissented from mainline hereditarianism.

In 1910, the EES pressed for legislation stalling the progeny of the feeble-minded, inebriates, carriers of venereal disease and other degenerate categories. The 'Feeble-Minded Persons Bill' was defeated in the House of Commons in 1911, but paved the way for the Mental Deficiency Act passed in 1914 sanctioning segregation of the feeble-minded and moral defectives in hospitals and institutions.[52] When the Act passed, Winston Churchill, Arthur Balfour and Neville Chamberlain had formally joined the EES, and under the Act that remained in place until 1959, thousands of people were incarcerated, including women regarded as morally degenerate.[53]

In 1929, the EES went a step further forming a Committee for Legalising Sterilisation that included Fisher, Huxley and Carr-Saunders. Backed by hefty scientific credentials, the proposals of the Committee received support from Sir Lawrence Brock appointed by the Ministry. Caradog Jones was another main player who defined the unfit as 'sub-normal types in the community, families containing individuals suffering from some defect, congenital or acquired, such as deafness, blindness or mental deficiency ... persons who had been in the receit of Public Assistance over a long period'.[54]

The Mental Deficiency Act had enshrined the 'problem group' as a category of mentally defective people defined on grounds of 'social incompetence', a speculative criterion denounced by geneticists like Hogben, Haldane and Penrose. Seeking to strengthen the scientific argument, Cora Hodson, the EES general secretary, approached Ernst Rüdin to inquire about the sterilization programme that underlay pre-Nazi legislation and overtly targeted social and pathological groups of 'abnormality'.[55] But opposition gathered and despite divisions, leading figures of the Royal College of Physicians, increasingly annoyed by eugenics ideology at odds with the ethos of their practice, reinforced the coalition against sterilization.[56]

John Macnicol explained how opposition to eugenic enactment came from several quarters.[57] The campaign, buttressed with inflammatory rhetoric to compensate for lack of data was a verdict, Macnicol suggested, that 'could well have been applied to the eugenics movement as a whole'.[58] The president of the Medico-Legal Society, Lord Ridell, alerted Chamberlain to the risk of lunatics and mental defectives taking over society, for such creatures were extremely persistent and clever. The sharpest blow against sterilization was delivered from the Fabian fringes of Labour, especially in the person of Hyacinth Morgan who pointed to 'economic stress, ill-nutrition, under-nourishment, bad midwifery, and acute microbic disease as the chief causes of deficiency, rather than the germ plasm'.[59]

Eugenic Liaisons: Sex, Contraception and Women

By framing reproduction and sexual matters as topics of public debate with social consequences eugenics also facilitated something more positive: widespread interest in sexual practices to be addressed in scientific publications, tracts, newspapers and political speeches. However the attitude of eugenicists in regard to sex, as with all matters to do with reproduction, was multifarious and fraught with double standards. Prominent women, intellectuals and activists viewed themselves as caretakers of the race and seized on the possibilities that eugenics offered to take control over their reproduction. Contraception was central to the debate that swirled around sexual mores.

Women were the reproducers and custodians of national fitness and the main object of positive and negative eugenic policies. But the marker of female unfitness through the lenses of female eugenicists could be as arbitrary as height, 'any woman less than five foot high was deemed dysgenic for after all: "No one can wish to perpetuate a race of dwarfs"'.[60] Motherhood among the fit was the cornerstone of the eugenic positive policy. As pioneering sexologist and leading EES officer Havelock Ellis put it: 'Women's function in life can never be the same as man's, if only because women are the mothers of the race ... the most vital problem before our civilisation today is the problem of motherhood, the question of creating human beings best suited for modern life.'[61]

Ellis lauded birth control as characteristic of the European and American races that disengaged sex from reproduction but his case is quite revealing. He was a fervent supporter of mainline eugenic tenets that could not be properly reconciled with his ideas on sexuality. In *Sexual Inversion* (1897) the discourse on the universal nature of homosexuality was suffused with the terminology of 'lower races' versus Western culture, and hypotheses about the passing of homosexual traits to children.

However, while homosexuality was not included in the 'category problem', it still involved mental pathologies and deviant social behaviour. Elaborating from a case witnessed in Switzerland, Ellis emphasized that 'Sterilisation by castration, offered a solution which was eagerly accepted by all parties'.[62] He then brought the lesson home: 'when we are dealing with the unfit the resources of civilisation in this matter are limited. And if we reject the sterilisation, what, I ask myself, is the practical alternative?'[63]

Like Shaw, Ellis was barred from the progressive club of Pearson. As Daniel Kevles suggested, the eugenics movement 'was Karl Pearson's Men and Women's Club – with its determination to explore the relations between the sexes – enlarged to encompass the transatlantic educated community ... like the members of the club, eugenicists divided on pertinent issues, particularly those rooted in sexuality'.[64] The club discussed topics like suffrage, prostitution, venereal disease, contraceptive methods, marriage and sexuality.

Mona Caird, a novelist representative of the ideals of the *New Woman*, was a member of Pearson's Club that perceptively captured the incongruity of the eugenic evolutionary discourse that tapped on jargon such as 'nature desires' or 'nature intends'. Darwinism, as Caird emphasized, did not entail biological determinism but stressed variation and eschewed teleology. She challenged the idea that evolutionary theory validated the status of women as anabolic creatures destined to evolutionary stasis and construed a plea for individual autonomy based on the teachings of John Stuart Mill.[65]

In contrast, the reformer Alice Ravenhill stressed 'no lack of eugenic ideals for womanhood' for 'what women would not aspire to advance racial progress'.[66] Such ideals, however, were not necessarily synonymous with emancipation. As an EES lecturer made plain: 'as Geddes and Thomson pointed out, the male is naturally active or katabolic and the female is passive or anabolic ... to foist excessive katabolic activity on an anabolic organism is not only unscientific but may be fraught with possible disaster'.[67] Feminist eugenicists caught in the grips of racial duty were lured by the prospect 'that women as sexual selectors could and should be to the fore in evolutionary development'.[68] But while eugenics offered women possibilities they had previously been denied, it did so in ways retaining the ascendance of male over female sexuality.

The EES was particularly concerned with matters of sexual education and hygiene and keen to gain better insights into the biological foundations of reproduction and sex. Birth control, however, was a minefield best avoided. Yet in the mind of eugenicists such as Ellis and Stopes there was a close link between the preventive checks of birth control and sterilization. After the defeat of the voluntary sterilization campaign, the EES proceeded to discuss contraceptive methods 'necessary in the interest, not only of the individual, but of the nation and the whole world to lessen the struggle for survival and the evils which result from it'.[69] In 1930, the National Birth Control Council (NBCC) congregated and although the EES did not formally join, its officers Julian Huxley, John Maynard Keynes, Leonard Darwin and Marie Stopes accepted positions in the NBCC.[70]

The task of enlightening the general public about the pleasures of sex in marriage and contraceptive methods was taken up by Marie Stopes with her best-sellers *Mother England* (1929) lauded by H. G. Wells as a striking and useful book, *Married Love* (1918) derided by Dora and Bertrand Russell as sentimental on the verge of comic, and *Wise Parenthood* (1918). While offering erotic instruction without crossing the boundaries of taste or decency, Stopes clung to traditional sexual patterns sanctioned by religion within marriage to the benefit of the race. The public, avid for information, reacted with a massive amount of letters asking for advice.

Stopes opened the first family planning clinic in Britain and advanced a cause calculated to appeal both to doctors and politicians. Speaking about the achievements of the Society for Constructive Birth Control and Racial Progress in Britain, Stopes observed that a 'sister Society in America, the Voluntary Parenthood League has a heavier task than ours ... to affect a change of Federal Law'.[71]

Eugenics in America: The Conundrum of 'Race Suicide' and Fertility

Galton's 1904 address to the Sociological Society was printed in both sides of the Atlantic, including H. G. Wells's claim to the effect that it is 'in the sterilization of failures, and not in the selection of success for breeding, that the possibility of an improvement of the human stock lies'.[72] In a world dominated by the anxieties of the white middle and upper classes woeful of demographic changes due to immigration, and thoroughly shaped by 'race suicide' concerns steered by intellectuals like R. R. Rentoul admonishing that 'hour by hour we add diseased humanity – the children begotten by the diseased, idiots, the imbeciles, epileptics, the insane', eugenics fitted rather well.[73]

The term 'race suicide' was coined by the sociologist Edward Ross, a founding member of the American Economic Association, eager to protect the 'native' Anglo-Saxon stock that risked being outbred by racially inferior immigrant races, 'Latins, Slavs, Asiatics, and Hebrews'.[74] In the heyday of the Progressive Era, Ross and Rentoul were speaking for many of their compatriots. Eugenics was rubber stamped when President Roosevelt addressed the nation about the dangers of race suicide.[75]

In the United States, eugenics was fostered through a powerful symbiotic link between politics and academics.[76] As Christine Rosen has pointed out, eugenics was vested with a sense of urgency and justified by science: 'correction could no longer be left to the ad hoc efforts of amateurs ... Eugenicists viewed the scourges of their age-pauperism, crime, disease, prostitution, alcoholism – not as evidence of individual moral failing, but as problems to be solved scientifically'.[77]

Geneticists, economists, demographers and statisticians forged a nexus of eugenic ideas enacted through public policy and legislation bearing on immigration, birth control and sterilization. Charles Davenport, who had visited Galton and Pearson in England, was the driving force beyond eugenics research by establishing a station for the experimental study of evolution at Cold Spring Harbor, funded by the Carnegie Institute. Like his British counterparts, Davenport used the pedigree methodology and identified good stock with the white middle classes, especially intellectuals, musicians and scientists. But for the construction of the 'category problem' he drew on Ross's bundle of immigrants to which he added the category of feeblemindedness despite uncertainty about its meaning. Goddard provided Davenport with crucial arguments for the eugenic cause with his pedigree study of the 'Kallikak', laying out the hereditary 'immoral tendencies' of moron women, criminals and other feebleminded people as a case of race suicide in impressionistic terms that appealed to his audience of the Progressive Era.[78]

Crucial for the foray into policy-making was Davenport's assistant, Harry Laughlin, who supervised the Eugenics Record Office. Laughlin, as Kevles pointed out, became known in Washington as an authority on immigration. The House Committee on Immigration and Naturalization was heavily reliant on research conducted in Cold Springs to substantiate the idea that recent immigrants and other defectives carried hereditary traits that undermined the national gene pool.[79] Their objective was twofold: restricting immigration and segregation of individuals in the 'category problem' in prisons, hospitals and charitable institutions.

The initial enthusiasm that geneticists demonstrated for eugenics was instrumental for its credibility among the segment of population whose interests it served, and the legislative triumphs achieved by the movement in the United

States. In the early years, biologists were keen to weigh whether genetic principles should form the basis of social legislation.[80] However, in the 1920s geneticists began expressing scepticism that genes could be solely responsible for complex traits and acknowledged that controlled breeding had little effect on the hereditary composition of the population. By the late 1930s, the most effective opposition to eugenics came from the 1922 American and British geneticists' *Manifesto*.[81]

In 1922, as a result of the backlash, the memorandum of the 'Second International Congress of Eugenics' states as a key objective the elimination of 'criticism of biologists in general, and of experimental geneticists in particular, to the effect that eugenicists knew too little of the foundation of inner science, and were engaged in the creation of a superstructure without sufficient attention to a firm base'.[82]

Economists contributed to back the argument on immigration through the amalgamation of evolutionary ideas and Malthusianism. While British eugenicists framed the differential in fertility primarily as an internal question, Americans fretted about the fertility ratio in relation to immigration. For the presidential address before the American Economic Association in 1913, Frank Fetter drew on Malthusianism to underpin a discourse on immigration inflected by eugenics. Irving Fisher, a firm believer in science with social implications, kept to the same line of argument buttressed with references to Galton, Davenport and a host of geneticists. According to his bleak diagnosis, 'statistics of the feeble-minded, insane, criminals, epileptics, inebriates, diseased, blind, deaf, deformed and dependent classes are not reassuring'.[83]

Fisher's corollary was plainspoken: 'If we allow ourselves to be a dumping ground for relieving Europe of its burden of defectives, delinquents, and dependants, while such action might be said to be humane for the present generation, it would be quite the contrary to the interests of humanity in the future.'[84] The slippage from the typical 'category problem' to a wider category covering European immigrants is quite revealing. As Thomas Leonard has argued, despite the decline in fertility characteristic of the Depression Era, the immigration to America of eastern and southern Europeans deemed racially inferior was 'effectively terminated by eugenic-inspired immigration restrictions, notably the Emergency Quota Act of 1921 and the Immigration Act of 1924'.[85]

As in the case of immigration, eugenic arguments in favour of sterilization intensified during the Depression. Eugenics was enacted on a large scale with laws of compulsory sterilization bearing on the 'category problem' overlapping race and degeneracy covering social and mental traits, with prostitution conspicuously linked to feeblemindedness. In 1907, the concerted efforts of Davenport, Goddard, Laughlin and others led to the Indiana enactment of the world's first involuntary sterilization law. The case of Buck versus Bell that allowed the State of Virginia to sterilize in 1927 an institutionalized, allegedly feebleminded single mother and her sister, through a decision by Justice Oliver Wendell, who befriended the Laskis, remains to the present day a case quoted in debates over reproductive rights. Neither of the women were feebleminded and the criterion for degeneracy was single motherhood resulting from incest. The Virginia law remained in place from 1924 to 1972 and at least 7,500 people in the state were sterilized. The California law, however, enabled the sterilization of more people than any other state, almost twice as much as the other states combined, a figure of 6,255 people by 1929.[86]

Women, as Wendy Kline has emphasized, were the first targeted, because eugenics had initially segregated 'moron girls' who challenged nineteenth-century notions of virtue, then switched to compulsory sterilization as a more effective and affordable solution to female sexual delinquency. Sterilization as a means of preventing procreation in light of human betterment appealed to a popular audience because it offered an efficient, scientific solution to the problem of racial degeneracy.[87]

Yet in the United States, like in Britain, eugenics found a large constituency of support among women. The American counterpart of Stopes was Margaret Sanger, the nurse and social reformer, who founded Planned Parenthood and helped generate shifts of opinion on issues that had not been previously negotiated in the public sphere by skilfully using eugenics as a forum to discuss family planning and sexual education.[88]

The mentor of the social warrior of the century, as publicity agents described Sanger, was Havelock Ellis: 'He, beyond any other person, has been able to clarify the question of sex, and free it from the smudginess connected with it.'[89] Ellis advised Sanger on the best strategy to make a compelling scientific case for contraception among the upper class and workers. In 1921, converted to Ellis's eugenic ideas, Sanger founded the American Birth Control League and a clinic. Partial to the healthy that bore the costs of those who should never have been born, Sanger was adamant that: 'eugenics without birth control seemed to me a house built upon sands ... to stop the multiplication of the unfit. This appeared the most important and greatest step towards race betterment.'[90]

In the United States, Sanger pondered, overpopulation was not a problem but quality became an issue so 'immigration laws ... barred aliens with mental, physical, communicable, or loathsome diseases, and also illiterate paupers, prostitutes, criminals and the feeble-minded. Had these precautions been taken earlier our institutions would not be crowded with moronic mothers, daughters ... detrimental to the blood stream of the race.'[91] Eugenics produced incommensurably deeper outcomes by way of policy and legislation in the United States than in Galton's native country, with wide-ranging implications for family planning and the sexual emancipation of women. However, legislation that enhanced women's control over reproduction could also be used to curtail their fertility and reproductive rights.

Eugenics Nowadays: Sneaking through the 'Backdoor'?

Rife with prejudice and entwined with the fate of evolutionary theory, eugenics, which contributed to scientific developments in many fields, particularly genetics, was by no means a 'fringe movement populated by a few zealots and pseudoscientists'.[92] As P. Fara put it, far 'from being a blind alley, eugenics is buried deep within the modern life sciences'.[93] It was, however, a treacherous alley. As Francis Crick and James Watson opened the door to the genomic era, their scientific breakthrough triggered concerns over the dawn of a second eugenics movement.

Scholars, practitioners and lay public understood that the benefits yielded by biotechnology implied added responsibilities. Troy Duster famously warned that 'when eugenics reincarnates this time, it will not come through the front door'.[94] Others pointed to the shortcomings of bioethical debates staged in a vacuum oblivious of the history of eugenics and confined to 'monmedical and nongenetical journals, so they are not readily available to geneticists or physicians'.[95]

Watson explained the implementation of ethical, legal and social issues research into genetic information (ELSI) to shield the Human Genome Project (HGP) on grounds that 'not forming a genome ethics program quickly might be falsely used as evidence that I was a closet eugenicist, having as my real long-term purpose the unambiguous identification of genes that lead to social and occupational stratification as well as genes justifying racial discrimination'.[96]

The engineering ambitions of eugenics live on in contemporary reproductive and biotech battles over policies overtly designated as eugenic, or unequivocally eugenic in substance regardless of denominations. Throughout history societies have appraised the fertility of individuals in terms of the benefits they yield to the community, and eugenics was just one of such interpretations despite the inequity it created. Old eugenics occasioned an understanding of the human condition conflicting with values of the Enlightenment legacy – individual autonomy, human life as end rather than means, pluralism, toleration and consent.

Eugenic proposals of the so-called liberal or mild guise have been much debated since the inception of the HGP. Abusive policies of rational reproduction and the misuse of genetic information for eugenic engineering that challenged freedom, personhood and individual autonomy in the past cast suspicion today over new reproductive technologies. As John Robertson put it, 'shaping or engineering offspring traits seems to be an unprecedented exercise of control over the lives of others ... may discriminate against women and the disabled, and threatens harm to offspring who are selected according to genetic criteria'.[97]

The fact that eugenics is today fostered less by governments than by individual decisions may suggest that former considerations do not apply. The scientific reification of group genetic identity was dispelled by population genetics. But national interests were replaced by the concept of common good conveying the idea that the individual, and his reproductive choices, are subservient to the public good. How to reconcile individual autonomy and rights with collective interests is an open question in open societies. Aviad Raz has recently suggested that eugenics can be redeemed in the context of reproductive genetics, for the distinctive focus of liberal eugenics is 'the individual, not the population, nation, race or class, and it gives primacy to the individuals own values and conceptions of what constitutes a good life'.[98]

Duster, on the other hand, has denounced liberal eugenics premised on individual choice as an illusion because individual autonomy is always relative so the prejudice of old eugenics lurking through the backdoor threatening individual rights. Among other practices, Duster exposed the contrast between mass health screening and information collected under the aegis of current genetic screening in the United States, supposed to determine 'risk populations' which, he claims, are inflected with standards of ethnicity and race so 'the *machinery in place* (organizational, institutional, legal, and physical) which will slowly ... shift the refraction of human traits, characteristics, behaviours, disorders, and defects through a "genetic prism"'.[99]

Boundaries between public and individual spheres are forever fungible. The old eugenic credo in social worth determined by genetic worth ranking individuals by fitness has not waned. As Sandra Peart and David Levy have emphasized, eugenics breached the postulate of homogeneity and individualism favoured by the Scottish Enlightenment given that in 'eugenic science, "experts" presupposed its subjects to be inferior and proposed to remake the human herd more to the experts' liking, to obtain racial perfection or for the "general good"'.[100]

The arrogance of expertise exposed by Peart and Levy pervades current debates on updated eugenics. F. Fukuyama has expressed confidence that state-sponsored eugenics is not likely to carry much weight in the future, and speaks of a kinder eugenics as a matter of individual choice on the part of parents, while casting a gloomy verdict on the effects of biotechnology.[101] As a cautionary tale, he cites the individualist engineering outlook defended by Ronald Dworkin, who 'proposes what amounts to a right to genetically engineer people, not so much on the part of parents but of scientists'.[102]

A notorious advocate of mild eugenics is Philip Kitcher who invokes as role models 'George Bernard Shaw, Sidney and Beatrice Webb, and others who followed them'.[103] By the same token, Kitcher has attempted to distance himself from old eugenics: 'the tendency to try to transform the population in a particular direction, not to avoid suffering but to reflect a set of social values'.[104] The mild eugenic outlook he advocates is underpinned by quality of life involving a threefold potential gauge: for self-determination, for central wishes to be fulfilled, for pleasurable experience. Gregory Radick set forth a compelling critique of Kitcher's argument by stressing that 'capacities for quality of life, and the consequences on others of a potential life, can never be exact'.[105]

Over the last three decades eugenics has become a watchword in the complex scientific/political debate over two interrelated issues: what biotech and reproductive genetic services are up to, and how to safeguard reproductive rights.[106] Ruth Chadwick's work on genetic counselling and Robert Sandel's ethics of genetic engineering represent opposing viewpoints in this debate. John Harris is a strong advocate of gene therapy who focuses on the elusive boundaries between enhancement and therapy and denounces Chadwick for arguing, on Galton's behalf, that 'those who are genetically weak should simply be discouraged from reproducing either by incentives or compulsory measures'.[107]

A. Buchanan, D. W. Brock, N. Daniels and D. Wilker argue for genetic intervention geared towards the personal choice model. Old eugenics, they claim, was not altogether flawed so new eugenics should not be eschewed unless it assumes the consequentialist outlook which not only allows but in some circumstances requires that the most fundamental interests of the individual be sacrificed in order to produce the best overall outcome.[108]

Consider those with a genetic defect who believe theirs is a life worth living. Do they embody a modern 'category problem'? There is much suspicion that eugenic proposals based on fuzzy assumptions about quality of life will pave the way for the arbitrary encroachment on individual autonomy. This suspicion has led Alasdair Palmer, himself a sufferer of multiple sclerosis, to call the results of genetic tests always to be balanced against the assertion that: 'I certainly think I have had, and still have, a life worth living'.[109]

Scholars are divided on these matters but many share the attitude of Michael Reiss, Roger Straughan, and Yoel Hashiloni-Dolev who prefer to exclude eugenics from contemporary debates. Eugenics, they stress, is a hazy category contaminated by its history, an 'emotive term' in the way of open discussions on the implications of biotech intervention. [110] Yet, the argument of Hashiloni-Dolev, underpinned by the dichotomy of a life worth/unworthy of living to spare suffering in the lives to come, calls up too clearly the spectre of eugenics and brings to light the scientific, legal, moral and political challenges raised by reproductive genetics.[111]

Conclusion

Eugenics emerged in the context of Victorian debates on the significance of evolution and incorporated the ideology of biological determinism. It assumed the inequality of human beings as a consequence of heredity and argued that particular categories of people weakened the gene pool. The movement sought to enhance the human species through rational reproduction but its most tangible outcome was policy aimed at eliminating unfit groups in society.

Current debates on eugenic intervention involve the issues of old eugenics: the benchmark for reproductive worth; hierarchy based on genetic endowment; and the issue of how society perceives and responds to 'fitness' and 'unfitness'. Each individual makes a distinct contribution, both positive and negative, to the biological makeup of the community but in open societies this unique contribution cannot be ranked. Genomic knowledge opened wide the doors for genetic engineering. However, the nature of desirable intervention if any remains unsolved.

NOTES

1 Reznick, 'From Birth to Death: The Evolution of Life Histories in Guppies (*Poecilia reticulata*)'

1. C. Darwin, *On the Origin of Species by Means of Natural Selection*, 1st edn (London: John Murray, 1859), p. 118.
2. R. A. Fisher, *The Genetical Theory of Natural Selection* (Oxford: Clarendon Press, 1930).
3. G. C. Williams, 'Natural Seletion, the Cost of Reproduction and a Refinement of Lack's Principle', *American Naturalist*, 100 (1966), pp. 687–90.
4. M. Gadgil and P. W. Bossert, 'Life Historical Consequences of Natural Selection', *American Naturalist*, 104 (1970), pp. 1–24.
5. R. Law, 'Optimal Life Histories under Age-Specific Predation', *American Naturalist*, 114 (1979), pp. 399–417; R. E. Michod, 'Evolution of Life Histories in Response to Age-Specific Mortality Factors', *American Naturalist*, 113 (1979), pp. 531–50; B. Charlesworth, *Evolution in Age Structured Populations* (Cambridge: Cambridge University Press, 1980).
6. C. P. Haskins, E. G. Haskins, J. J. McLaughlin and R. E. Hewitt, 'Polymorphism and Population Structure in *Lebistes reticulata*, a Population Study', in W. F. Blair (ed.), *Vertebrate Speciation* (Austin, TX: University of Texas Press, 1961), pp. 320–95; J. A. Endler, 'A Predator's View of Animal Color Patterns', *Evolutionary Biology*, 11 (1978), pp. 319–64.
7. Haskins et al., 'Polymorphism and Population Structure'; Endler, 'A Predator's View of Animal Color Patterns'.
8. D. N. Reznick, M. J. Butler IV, F. H. Rodd and P. Ross, 'Life History Evolution in Guppies (*Poecilia reticulata*). 6. Differential Mortality as a Mechanism for Natural Selection', *Evolution*, 50 (1996), pp. 1651–60.
9. H. J. Alexander, J. S. Taylor, S. S. T. Wu and F. Breden, 'Parallel Evolution and Vicariance in the Guppy (*Poecilia reticulata*) over Multiple Spatial and Temporal Scales', *Evolution*, 60 (2006), pp. 2352–69.
10. Endler, 'A Predator's View of Animal Color Patterns'. J. A. Endler, 'Natural Selection on Color Patterns in *Poecilia reticulata*', *Evolution*, 34 (1980), pp. 76–91.
11. D. N. Reznick and H. Bryga, 'Life-History Evolution in Guppies. 1. Phenotypic and Genotypic Changes in an Introduction Experiment', *Evolution*, 41 (1987), pp. 1370–85. D. Reznick, H. Bryga and J. A. Endler, 'Experimentally Induced Life-History Evolution in a Natural Population', *Nature*, 346 (1990), pp. 357–9. D. N. Reznick, F. H. Shaw, R. H. Rodd and R. G. Shaw, 'Evaluation of the Rate of Evolution in Natural Populations of Guppies (*Poecilia reticulata*)', *Science*, 275 (1997), pp. 1934–7.

12. D. N. Reznick, M. J. Butler IV and F. H. Rodd, 'Life History Evolution in Guppies 7: The Comparative Ecology of High and Low Predation Environments', *American Naturalist*, 157 (2001), pp. 126–40.
13. C. C. Smith and S. D. Fretwell, 'Optimal Balance between Size and Number of Offspring', *American Naturalist*, 108 (1974), pp. 499–506.
14. C. Jorgensen, S. K. Auer and D. N. Reznick, 'A Model for Optimal Offspring Size in Fish, including Live-Bearing and Parental Effects', *American Naturalist*, 177 (2011), pp. E119–E135.
15. Jorgensen et al., 'A Model for Optimal Offspring Size'.
16. F. Bashey, 'Competition as a Selective Mechanism for Larger Offspring Size in Guppies', *Oikos*, 117 (2008).
17. G. C. Williams, 'Pleiotropy, Natural Selection and the Evolution of Senescence', *Evolution*, 11 (1957), pp. 398–411.
18. P. B. Medawar, *An Unsolved Problem of Biology* (London: H. K. Lewis & Co. Ltd, 1952).
19. D. N. Reznick, M. J. Bryant, D. Roff, C. K. Ghalambor and D. E. Ghalambor, 'Effect of Extrinsic Mortality on the Evolution of Senescence in Guppies', *Nature*, 431 (2004), pp. 1095–9.
20. L. Partridge and N. H. Barton, 'On Measuring the Rate of Ageing', *Proceedings of the Royal Society of London, Series B*, 263 (1996), pp. 1365–71.
21. Reznick et al., 'Effect of Extrinsic Mortality'.
22. Williams, 'Pleiotrophy, Natural Selection and the Evolution of Senescence'.
23. Medawar, *An Unsolved Problem of Biology*.
24. Williams, 'Pleiotropy, Natural Selection and the Evolution of Senescence'.
25. Reznick et al., 2001.
26. D. N. Reznick, C. Ghalambor and L. Nunney, 'The Evolution of Senescence in Fish', *Mechanisms of Ageing and Development*, 123 (2002), pp. 773–89.
27. B. Charlesworth, *Evolution in Age Structured Populations* (Cambridge: Cambridge University Press, 1980); P. Abrams, 'Does Increased Mortality Favor the Evolution of More Rapid Senescence?', *Evolution*, 47 (1993), pp. 877–87.
28. Reznick et al., 'Life History Evolution in Guppies 7: The Comparative Ecology of High and Low Predation Environments'.
29. Abrams, 'Does Increased Mortality Favor the Evolution of More Rapid Senescence?'; P. D. Williams and T. Day, 'Antagonistic Pleiotropy, Mortality Source Interactions and the Evolutionary Theory of Senescence', *Evolution*, 57 (2003), pp. 1478–88.
30. Medawar, *An Unsolved Problem of Biology*.
31. Williams, 'Pleiotropy, Natural Selection and the Evolution of Senescence'.
32. Charlesworth, *Evolution in Age Structured Populations*; Abrams, 'Does Increased Mortality Favor the Evolution of More Rapid Senescence?'
33. J. L. Dudycha and A. J. Tessier, 'Natural Genetic Variation of Life Span, Reproduction, and Juvenile Growth in *Daphnia*', *Evolution*, 53 (1999), pp. 1744–56.
34. G. A. Wellborn, 'Size-Biased Predation and Prey Life Histories: A Comparative Study of Freshwater Amphipod Populations', *Ecology*, 75 (1994), pp. 2104–17.
35. M. R. Rose and B. Charlesworth, 'Genetics of Life History in *Drosophila melanogaster*. I. Sib Analysis of Adult Females', *Genetics*, 97 (1981), pp. 173–86; M. R. Rose and B. Charlesworth, 'Genetics of Life History in *Drosophila melanogaster*. II. Exploratory Selection Experiments', *Genetics*, 97 (1981), pp. 187–96; M. R. Rose, 'Laboratory Evolution of Postponed Senescence in *Drosophila melanogaster*', *Evolution*, 38 (1984), pp. 1004–10.

36. S. C. Stearns, M. Ackermann, M. Doebeli and M. Kaiser, 'Experimental Evolution of Aging, Growth, and Reproduction in Fruitflies', *Proceedings of the National Academy of Sciences of the United States of America*, 97 (2000), pp. 3309–13.
37. L. S. Luckinbill and M. J. Clare, 'A Density Threshold for the Expression of Longevity in *Drosophila melanogaster*', *Heredity*, 56 (1986), pp. 329–36.
38. Stearns et al., 'Experimental Evolution of Aging'.
39. D. N. Reznick and C. K. Ghalambor, 'Can Commercial Fishing Cause Evolution? Answers from Guppies (Poecilia reticulata)', *Canadian Journal of Fisheries and Aquatic Sciences*, 62 (2005), pp. 791–801.
40. G. Peron, O. Gimenez, A. Chamantier, J. M. Gaillard and P. A. Crochet, 'Age at Onset of Senescence in Birds and Mammals is Predicted by Early-Life Performance', *Proceedings of the Royal Society of London, Series B*, 277 (2010), pp. 2849–56.
41. Williams, 'Pleiotropy, Natural Selection and the Evolution of Senescence'.
42. See, for example, M. Power, 'Effects of Fish in River Food Webs', *Science*, 250 (1990), pp. 811–14; M. Power, 'Top-Down and Bottom-Up Forces in Food Webs: Do Plants have Primacy?', *Ecology*, 73 (1992), pp. 733–46; R. J. Marquis and C. J. Whelan, 'Insectivorous Birds Increase Growth of White Oak through Consumption of Leaf-Chewing Insects', *Ecology*, 75 (1994), pp. 2007–14; D. K. Letourneau and L. A. Dyer, 'Experimental Test in Lowland Tropical Forest Shows Top-Down Effects through Four Trophic Levels', *Ecology*, 79 (1998), pp. 1678–87; D. R. Strong, A. V. Whipple, A. L. Child and B. Dennis, 'Model Selection for a Subterranean Tropchi Cascade: Root-Feeding Catepillars and Entomopathogenic Nematodes', *Ecology*, 80 (1999), pp. 2750–61; E. L. Preisser, 'Field Evidence for a Rapidly Cascading Underground Food Web', *Ecology*, 84 (2003), pp. 869–74.
43. R. T. Paine, 'Food Web Complexity and Species Diversity', *American Naturalist*, 100 (1966), pp. 65–75; R. T. Paine, 'A Note on Trophic Complexity and Community Stability', *American Naturalist*, 103 (1969), pp. 91–3.
44. L. S. Mills, M. E. Soule and D. F. Doake, 'The Keystone-Species Concept in Ecology and Conservation', *BioScience*, 43 (1993), pp. 219–23.
45. J. P. Wright, C. G. Jones and A. S. Flecker, 'An Ecosystem Engineer, the Beaver, Increases Species Richness at the Landscape Scale', *Oecologia*, 132 (2002), pp. 96–101.
46. A. S. Flecker, 'Ecosystem Engineering by a Dominant Detritivore in a Diverse Tropical Stream', *Ecology*, 77 (1996), pp. 1845–54.
47. W. H. Romme, M. G. Turner, L. L. Wallace and J. S. Walker, 'Aspen, Elk and Fire in Northern Yellowstone Park', *Ecology*, 76 (1995), pp. 2097–106; W. J. Ripple and E. J. Larsen, 'Historic Aspen Recruitment, Elk, and Wolves in the Northern Yellowstone National Park, USA', *Biological Conservation*, 95 (2000), pp. 361–70; W. J. Ripple, E. J. Larsen, R. A. Renkin and D. W. Smith, 'Trophic Cascades among Wolves, Elk and Aspen on Yellowstone National Parl's Northern Range', *Biological Conservation*, 102 (2001), pp. 227–34; D. W. Smith, R. O. Peterson and D. B. Houston, 'Yellowstone after Wolves', *BioScience*, 53 (2003), pp. 330–40; D. Fortin, H. L. Beyer, M. S. Boyce, D.W. Smith, T. Duchesne and J.S. Mao, 'Wolves Influence Elk Movements: Behavior Shapes a Trophic Cascade in Yellowstone National Park', *Ecology*, 86 (2005), pp. 1320–30.
48. See Smith et al., 'Yellowstone after Wolves'; Fortin et al., 'Wolves Influence Elk Movements'.
49. J. A. Estes, J. Terborgh, J. S. Brashares, M. E. Power, J. Berger, W. J. Bond, S. R. Carpenter, T. E. Essington, R. D. Holt, J. B. C. Jackson, R. J. Marquis, L. Oksanen, T. Oksanen, R. T. Paine, E. K. Pikitch, W. J. Ripple, S. A. Sandin, M. Scheffer, T. W. Schoener, J. B. Shurin, A. R. E. Sinclair, M. E. Soule.

2 De Sousa, 'If we have Sex, do we have to Die?'

1. L. Angel, *Enlightenment East and West* (SUNY Press, 1994).
2. J. Moussaieff Masson, *When Elephants Weep: The Emotional Lives of Animals* (New York: Delacorte Press, 1995) and R. Hooper, 'Death in Dolphins: Do they Understand they are Mortal?', *New Scientist*, 2828 (1 September 2011).
3. See T. Aquinas, *The Summa Theologica of St. Thomas Aquinas, Second and Revised Edition,* ed. K. Knight (1920; New Advent, 2008, II-2-Q154), and I. Kant, 'Duties Towards the Body in Respect of the Sexual Impulse', in *Lectures on Ethics*, trans. L. Infield (Cambridge: Cambridge University Press, 1997), pp. 155–61.
4. M. Zuk, *Sex on Six Legs: Lessons on Life, Love, and Language from the Insect World* (New York: Houghton Mifflin Harcourt, 2011).
5. J. Ackerman, 'Dragonflies Strange Love', *National Geographic*, 209:4 (2006), pp. 104–7, on p. 104.
6. According to Pascal Boyer, a crucial condition for an idea to be retained and disseminated is that it should contradict an evident truism, while also exploiting familiar and well-established categories. The immortality of the soul, for example is made all the more attractive by contradicting the obvious facts of biology. See P. Boyer, *Religion Explained: The Evolutionary Origins of Religious Thought* (New York: Basic Books, 2001).
7. J. Maynard Smith, *The Evolution of Sex* (Cambridge: Cambridge University Press, 1978).
8. J. C. Cole, 'Unisexual Lizards', *Scientific American,* 250 (1984), pp. 94–100.
9. Aristotle, *Metaphysics*, 6 vi, in *The Complete Works of Aristotle: The Revised Oxford Translation*, ed. J. Barnes, Bollingen Series (Princeton, NJ: Princeton University Press, 1984).
10. See R. Millikan, *Language, Thought, and other Biological Categories* (Cambridge, MA: MIT Press, 1984), p. 34. It is possible, though not well established, that this must be qualified, if we believe the hypothesis put forward by R. Baker and M. Bellis in *Human Sperm Competition: Copulation, Masturbation and Infidelity* (London: Chapman Hall, 1995), according to which most spermatozoa actually have the function of killing other spermatozoa from a rival male, pp. 259–60.
11. For an excellent collection of papers on objections and proposed refinements of the aetiological conception, see C. Allen, M. Bekoff and G. Lauder, *Nature's Purposes: Analyses of Function and Design in Biology* (Cambridge, MA: MIT Press, 1998).
12. B. Hölldobler and E. O. Wilson, *The Superorganism: The Beauty, Elegance, and Strangeness of Insect Societies* (New York: Norton, 2008).
13. R. Dawkins, *The Extended Phenotype: The Gene as Unit of Selection* (Oxford: Oxford University Press, 1982).
14. S. Oyama, 'Causal Democracy and Causal Contributions in Developmental Systems Theory', *Philosophy of Science,* 67 (2000), pp. S332–S347.
15. E. Jablonka, 'Information: its Interpretation, its Inheritance, and its Sharing', *Philosophy of Science,* 69 (2002), pp. 578–605.
16. P. Teilhard de Chardin, *The Human Phenomenon,* trans. S. Appleton-Weber (Hove: Sussex Academic Press, 2003).
17. For a comprehensive survey of the controversy, see R. C. Pennock (ed.), *Intelligent Design Creationism and Its Critics: Philosophical, Theological and Scientific Perspectives* (Cambridge, MA: MIT Press, 2001).
18. M. J. Behe, *Darwin's Black Box: The Biochemical Challenge to Evolution* (New York: Simon & Schuster, 1998).
19. J. Ruffié, *Le Sexe et la Mort* (Paris: Odile Jacob, 1986), p. 21.

20. Some biologists have argued, however, that a broad trend towards greater complexity follows from the simple fact that in a random walk there is an absolute level of minimal complexity compatible with life, but no absolute maximum. Certain specific 'discoveries' in evolution, such as eyes or wings, have evolved several times independently, which suggests that certain outcomes might be highly probable in the long run, even if these are attained by different paths. Simon Conway Morris has argued that the existence of numerous cases of convergent evolution shows that intelligent human-like beings were bound to arise sooner or later. See S. Conway Morris, *Life's Solution: Inevitable Humans in a Lonely Universe* (New York: Cambridge University Press, 2003).
21. M. Ridley, *Mendel's Demon: Gene Justice and the Complexity of Life* (London: Weidenfeld, 2000).
22. Estimates in these matters differ by an order of magnitude. The figures cited here were retrieved from http://www.wolframalpha.com [accessed 15 October 2011].
23. C. Greider, 'Telomerase Discovery: The Excitement of Putting Together Pieces of the Puzzle' (Nobel Prize Lecture (2009), at http://www.nobelprize.org/mediaplayer/index.php?id=1216 [accessed 08 April 2012].
24. J. C. Ameisen, 'Carving Life: Regulated Self-Destruction and the Evolution of Complexity', presented at the Fire in the Crystal: The Mystery of Life in the Recurrent Geometric Patterns of the Universe. 27th International Conference, Rimini, Pio Manzù Research Centre, ONU, UNIDO 2002.
25. R. Skloot, *The Immortal Life of Henrietta Lacks* (New York: Crown, 2010).
26. In this section, I have made extensive use of William Clark's fascinating book, *Sex and the Origins of Death* (New York: Oxford University Press, 1996).
27. L. Margulis and D. Sagan, *Acquiring Genomes: A Theory of the Origins of Species* (New York: Basic, 2002).
28. F. Crick, 'Central Dogma of Molecular Biology', *Nature,* 227 (1970), pp. 561–3.
29. J. Barnes, *The Presocratic Philosophers* (New York: Routledge, 1983), p. 102.
30. E. Sober and D. S. Wilson, *Unto Others: The Evolution and Psychology of Unselfish Behavior* (Cambridge, MA: Harvard University Press, 1998), p. 350.
31. O. Blanke et al., 'Stimulating Illusory Own-Body Perceptions: The Part of the Brain that can Induce Out-of-Body Experiences has been Located', *Nature,* 419 (2002), p. 269.
32. W. Buss, *The Evolution of Individuality* (Princeton, NJ: Princeton University Press, 1987).
33. W. Hamilton, 'William Hamilton, Evolutionary Revolutionary', at http://web.archive.org/web/20070413182558/http://www.goodbyemag.com/mar00/hamilton.html [accessed 6 April 2012].
34. The protection is not absolute. It is well known that the gametes produced by older men are more likely to carry birth defects due to mutations, which suggests that the sequestration of the sex cell lineage does not protect them absolutely. (It also serves to remind us that most interesting facts in biology are statistical, not nomological.) I am grateful to Gonzalo Munévar for reminding me of this fact - as well as for a number of other corrections and suggestions to a draft of the present paper.

3 Kraaijeveld and Bast, 'The Genomic Consequences of Asexual Reproduction'

1. J. Maynard Smith, *The Evolution of Sex* (Cambridge: Cambridge University Press, 1978).
2. J. Mergeay, D. Verschuren and L. De Meester, 'Invasion of an Asexual American Water Flea Clone throughout Africa and Rapid Displacement of a Native Sibling Species', *Proc. R. Soc. Lond. B,* 273 (2006), pp. 2839–44.

3. G. Bell, *The Masterpiece of Nature* (Berkeley, CA: Unversity of California Press, 1982); I. Schön, K. Martens and P. van Dijk (eds), *Lost Sex* (Dordrecht: Springer, 2009).
4. For more complete summaries, see Bell, *The Masterpiece of Nature*, and Schön, et al. (eds), *Lost Sex*.
5. C. Moritz, 'Parthenogenesis in the Endemic Australian Lizard *Heteronotia binoei* (Gekkonidae)', *Science*, 220 (1983), pp. 735–7.
6. K. P. Lampert and M. Schartl, M, 'The Origin and Evolution of a Unisexual Hybrid: *Poecilia formosa*', *Phil. Trans. R. Soc. B*, 363 (2008), pp. 2901–09.
7. B. A. Pannebakker, L. P. Pijnacker, B. J. Zwaan and L. W. Beukeboom, 'Cytology of *Wolbachia*-induced Parthenogenesis in *Leptopilina clavipes* (Hymenoptera: Figitidae)', *Genome*, 47 (2004), pp. 299–303.
8. D. Fournier, A. Estoup, J. Orivel, J. Foucaud, H. Jourdan, J. Le Breton and L. Keller, 'Clonal Reproduction by Males and Females in the Little Fire Ant', *Nature*, 435 (2005), pp. 1230–4.
9. W. Rice, 'Experimental Tests of the Adaptive Significance of Sexual Recombination', *Nature Reviews Genetics*, 3 (2002), pp. 241–51.
10. M. R. Goddard, H. C. J. Godfray and A. Burt, 'Sex Increases the Efficacy of Natural Selection in Experimental Yeast Populations', *Nature*, 434 (2005), pp. 636–40.
11. S. Scheu and B. Drossel, 'Sexual Reproduction Prevails in a World of Structured Resources in Short Supply', *Proc. R. Soc. Lond. B*, 274 (2007), pp. 1225–31; Y. Song, B. Drossel and S. Scheu, 'Temporal Patterns of Resource Usage in an Ecological Model for Sexual Reproduction and Geographic Parthenogenesis', *Evol. Ecol. Res.*, 12 (2010), pp. 831–41.
12. A. S. Kondrashov, 'Deleterious Mutations and the Evolution of Sexual Reproduction', *Nature*, 336 (1988), pp. 435–40; J. F. Crow, 'Advantages of Sexual Reproduction', *Dev. Genet.*, 15 (1994), pp. 205–13.
13. L. T. Moran, M. D. Parmenter and P. C. Phillips, 'Mutation Load and Rapid Adaptation Favour Outcrossing over Self-fertilization', *Nature*, 462 (2009), pp. 350–2.
14. A. Burt and R.Trivers, *Genes in Conflict: the Biology of Selfish Genetic Elements*. (Cambridge: Belknapp Press, 2006).
15. T. Wicker, F. Sabot, A. Hua-Van, J. L. Bennetzen, P. Capy, B. Chalhoub, A. Flavell, P. Leroy, M. Morgante, O. Panaud, E. Paux, P. SanMiguel and A. H. Schulman, 'A Unified Classification System for Eukaryotic Transposable Elements', *Nat. Rev. Genet.*, 8 (2007), pp. 973–83.
16. J. Werren, 'Selfish Genetic Elements, Genetic Conflict and Evolutionary Innovation', *Proc. Natl. Acad. Sci. USA*, 108 (2011), pp. 10863–70.
17. E. S. Dolgin and B. Charlesworth, 'The Fate of Transposable Elements in Asexual Populations', *Genetics*, 174 (2006), pp. 817–27.
18. E. S. Dolgin, B. Charlesworth and A. D. Cutter, 'Population Frequencies of Transposable elements in Selfing and Outcrossing', *Caenorhabditis* Nematodes. *Genet. Res. Camb.*, 90 (2008), pp. 317–29; S. I. Wright, Q. H. Le, D. J. Schoen and T. E. Bureau, 'Populations Dynamics of an AC-like Transposable Element in Self- and Cross-Pollinating *Arabidopsis*', *Genetics*, 158 (2001), pp. 1279–88.
19. M. Rho, S. Schaack, X. Gao, S. Kim, M. Lynch and H. Tang, 'LTR Retroelements in the Genome of *Daphnia pulex*', *BMC Genomics*, 11 (2010), pp. 425; S. Schaack, E. Choi, M. Lynch and E. Pritham, 'DNA Transposons and the Role of Recombination in Mutation Accumulation in *Daphnia pulex*', *Genome Biology*, 11 (2010), pp. R46.

20. I. Arkhipova and M. Meselson, 'Transposable Elements in Sexual and Ancient Asexual Taxa', *Proc. Natl. Acad. Sci. USA*, 97 (2000), pp. 14473–7.
21. E. A. Gladyshev and I. R. Arkhipova, 'Genome Structure of Bdelloid Rotifers: Shaped by Asexuality or Dessication? *J. Heredity*, 101 (2010), pp. S85–S93.
22. E. C. Verhulst, L. W. Beukeboom and L. van der Zande, 'Maternal Control of Haplodiploid Sex Determination in the Wasp *Nasonia*, *Science*, 328 (2010), pp. 620–3.
23. N. Eckardt, 'A Role for ARGONAUTE in Apomixis', *The Plant Cell*, 23 (2011), pp. 430.
24. V. Olmedo-Monfil, N. Duran-Figueroa, M. Artgeaga-Vazquez, E. Demese-Arevalo, D. Autran, D. Grimanelli, R. K. Slotkin, R. A. Martienssen and J. P. Vielle-Calzada, 'Control of Female Gamete Formation by a Small RNA Pathway in *Arabidopsis*', *Nature*, 464 (2010), pp. 628–34.
25. O. P. Judson and B. B. Normark, 'Ancient Asexual Scandals', *Trends Ecol. Evol.*, 11 (1996), pp. A41–A46.
26. D. Mark Welch and M. Meselson, 'Evidence for the Evolution of Bdelloid Rotifers Without Sexual Reproduction or Genetic Exchange', *Science*, 288 (2000), pp. 1211–15.
27. E. A. Gladyshev, M. Meselson and I. R. Arkhipova, 'Massive Horizontal Gene Transfer in Bdelloid Rotifers', *Science*, 320 (2008), pp. 1210–13.

4 Hattiangadi, 'Evolution and Illusion'

1. A. J. Schopenhauer, *The World as Will and Representation,* trans. E. J. F. Payne, 2 vols (New York: Dover Publications, 1958.)
2. J. J. Gibson, *The Ecological Approach to Visual Perception* (Boston, MA: Houghton Mifflin, 1979.)
3. J. Maynard Smith, *The Evolution of Sex* (Cambridge: Cambridge University Press, 1972.)
4. Ibid.
5. M. Ghiselin, *The Economy of Nature and the Evolution of Sex* (Berkeley, CA: University of California Press, 1974.)
6. N. Eldredge and S. J. Gould, 'Punctuated Equilibria: An Alternative to Phyletic Gradualism', in T. J. M. Schopf (ed.), *Modern Paleobiology* (San Francisco, CA: Freeman, Copper & Company, 1972), pp. 82–115.
7. G. C. Williams , *Sex and Evolution* (Princeton, NJ: Princeton University Press, 1975.)
8. H. J. Muller, 'Some Genetic Aspects of Sex', *American Nauralist,* 8 (1932), pp. 118–38.
9. G. Bell, *The Masterpiece of Nature: The Evolution and Genetics of Sexuality*(London: Croom Helm, 1982.)
10. Williams, *Sex and Evolution*.
11. This account was proposed by Francis Bacon and developed by Robert Boyle and Isaac Newton, to be described in my forthcoming book on *The Theory and Craft of Breaking Through in Science.*
12. R. Dawkins, *The Selfish Gene* (New York and Oxford: Oxford University Press, 1976.)
13. L. Margulis, *Origin of Eukaryotic Cells* (New Haven, CT: Yale University Press, 1971) and earlier by her in L. Sagan, 'On the Origin of Mitosing Cells', *Journal of Theoretical Biology,* 14:3 (1967), pp. 255–74.
14. L. Margulis, *The Origin of Sex: Three Billion Years of Genetic Recombination* (New Haven, CT: Yale University Press, 1990.)

15. C. De Duve, *A Guided Tour of the Living Cell* (New York: Scientific American Books, W, H. Freeman & Company, 1984.)
16. Y. Davidov and E. Jurkevitch, 'Predation between Prokaryotes and the Origin of Eukaryotes', *BioEssays*, 31 (2009), pp. 748–57.
17. Ibid.
18. D. G. Stearcy, 'Origins of Mitochondria and Chloroplasts from Sulphur Based Symbioses', in H. Hartman and K. Matsumo (eds) *The Origin and Evolution of the Cell, Proceedings of Conference, April, 1992* (Singapore: World Scientific Publishing Company, 1992), pp. 47–78.
19. C. De Duve, 'Evolution of the Peroxisomes', *Annals of the New York Academy of Science*, 168 (1969), pp. 369–81; W. R. Martin and M. Müller, 'The Hydrogen Hypothesis for the First Eukaryote' *Nature*, 392 (1998), pp. 37–41.
20. W. R. Martin, *Philosophical Transactions of the Royal Society*, 365 (2010), pp. 847–55.

5 De Block and Newson, 'Evolutionary Theory, Constructivism and Male Homosexuality'

1. R. Dawkins, *River Out of Eden* (New York: Basic Books, 1995).
2. E. Stein, *The Mismeasure of Desire: The Science, Theory, and Ethics of Sexual Orientation* (Oxford: Oxford University Press, 1999), p. 124.
3. K. Laland and G. Brown, *Sense and Nonsense: Evolutionary Perspectives on Human Behaviour* (Oxford: Oxford University Press, 2011), p. 55.
4. D. S. Wilson, 'Evolutionary Social Constructivism', in J. Gottschall and D. S. Wilson (eds), *The Literary Animal: Evolution and the Nature of Narrative* (Evanston: Northwestern University Press, 2005), pp. 20–37; and A. De Block and B. Du Laing, 'Paving the Way for an Evolutionary Social Constructivism', *Biological Theory*, 2 (2007), pp. 403–12.
5. I. Hacking, *The Social Construction of What?* (Cambridge, MA: Harvard University Press), p. 3.
6. A. Lakoff, *Pharmaceutical Reason. Knowledge and Value in Global Psychiatry* (Cambridge: Cambridge University Press, 2005).
7. See for example D. M. Halperin, *One Hundred Years of Homosexuality and Other Essays on Greek Love* (New York: Routledge, 1990); D. M. Halperin, *How To Do the History of Homosexuality* (Chicago, IL: University of Chicago Press, 2002); and R. Trumbach, *Sex and the Gender Revolution, Volume Two: The Origins of Modern Homosexuality* (Chicago, IL: University of Chicago Press, forthcoming). For contrary opinions see for example K. Borris and G. Rousseau (eds), *The Sciences of Homosexuality in Early Modern Europe* (London: Routledge, 2008); J. Boswell, 'Revolutions, Universals, and Sexual Categories', in M. Duberman, M. Vicinus, and G. Chauncey (eds), *Hidden from History: Reclaiming the Gay and Lesbian Past* (New York: Penguin Books, 1989), pp. 17–36; and R. Norton, *The Myth of the Modern Homosexual: Queer History and the Search for Cultural Unity* (London: Cassell, 1997).
8. M. Foucault, *The History of Sexuality (Volume 1)* (New York: Random House, 1980), p. 43.
9. See for example Crompton, *Homosexuality and Civilization*; Murray, *Homosexualities*; Trumbach, *Sex and the Gender Revolution*.

10. D. J. Noordam, 'Sodomy in the Dutch Republic, 1600–1725', in K. Gerard and G. Hekma (eds), *The Pursuit of Sodomy: Male Homosexuality in Renaissance and Enlightenment Europe* (London: Haworth Press, 1989), pp. 207–28.
11. G. H. Herdt (ed.), *Ritualized Homosexuality in Melanesia* (Berkeley, CA: University of California Press, 1984).
12. See for example T. Van der Meer, 'The Persecutions of Sodomites in Eighteenth-Century Amsterdam: Changing Perceptions of Sodomy', in K. Gerard and G. Hekma (eds), *The Pursuit of Sodomy: Male Homosexuality in Renaissance and Enlightenment Europe* (London: Haworth Press, 1989), pp. 263–310; Murray, *Homosexualities*; and H. Oosterhuis, *Stepchildren of Nature: Krafft-Ebing, Psychiatry and the Making of Sexual Identity* (Chicago, IL: University of Chicago Press, 2000). Historical court records and penal codes provide us with an important argument for this claim. In pre-modern times, criminal prosecution of same-sex sexuality was often based solely on the question whether there had been actual sexual contact between partners, particularly anal penetration ('res in re et effusio seminis'). The feelings and desires of the defendants were deemed irrelevant. Reversely, modern law systems were eager to incorporate intentions in sentencing same-sex sexual partners, thus emphasizing the interiorization of sexual preference (see especially Van der Meer, 'The Persecutions of Sodomites in Eighteenth-Century Amsterdam').
13. M. McIntosh, 'The Homosexual Role', *Social Problems*, 16:2 (1968), pp. 182–92; and R. Trumbach, 'Gender and the Homosexual Role in Modern Western Culture: The 18th and 19th Centuries Compared', in D. Altman (ed), *Homosexuality, Which Homosexuality?* (London: GMP Publishers, 1989), pp. 149–69.
14. Murray, *Homosexualities*, p. 421.
15. M. Pigliucci, *Phenotypic Plasticity: Beyond Nature and Nurture* (Baltimore, MD: Johns Hopkins University Press, 2001).
16. Wilson, 'Evolutionary Social Constructivism', p. 23.
17. J. Belsky, 'Conditional and Alternative Reproductive Strategies: Individual Differences in Susceptibility to Rearing Experiences', in J. Rodgers, D. Rowe and W. Miller (eds), *Genetic Influences on Human Fertility and Sexuality: Theoretical and Empirical Contributions from the Biological and Behavioral Sciences* (Boston, MA: Kluwer Academic Publishing, 2000), pp. 127–46.
18. D. J. Barker, *Mothers, Babies and Health in Later Life* (Edinburgh: Churchill Livingstone, 1998).
19. Along with their genes organisms inherit regulators for these genes – essentially switches that determine which genes will be expressed and which will be ignored. Genes are turned on and off depending on whether their information is useful in the circumstances. During embryonic development, cells specialize and genes are permanently switched off as, for example, muscle cells pack away genes that need only be expressed in liver cells. See E. Jablonka and G. Raz, 'Transgenerational Epigenetic Inheritance: Prevalence, Mechanisms, and Implications for the Study of Heredity and Evolution', *Quarterly Review of Biology*, 84 (2009), pp. 131–76.
20. See for example R. C. Kirkpatrick, 'The Evolution of Human Homosexual Behavior', *Current Anthropology*, 41:3 (2000), pp. 385–98; F. Muscarella, 'The Evolution of Homoerotic Behavior in Humans', *Journal of Homosexuality*, 40 (2000), pp. 51–77; and P. L. Vasey and V. Sommer, 'Homosexual Behaviour in Animals: Topics, Hypotheses and Research Trajectories', in V. Sommer and P. L. Vasey (eds), *Homosexual Behaviour in Animals: An Evolutionary Perspective* (Cambridge: Cambridge University Press, 2006), pp. 3–44.

21. J. Mann, 'Establishing Trust: Socio-Sexual Behaviour and the Development of Male–Male Bonds among Indian Ocean Bottlenose Dolphins', in Sommer and Vasey (eds), *Homosexual Behaviour in Animals*, pp. 107–30; K. Kotrschal, J. Hemetsberger, and B. M. Weiß, 'Homosociality in Male Greylag Geese (*Anser anser*): Making the Best of a Bad Situation', in Sommer and Vasey (eds), *Homosexual Behaviour in Animals*, pp. 45–76; E. A. Fox, 'Homosexual Behavior in Wild Sumatran Orangutans (*Pongo pygmaeus abelii*)', *American Journal of Primatology*, 55:3 (2001), pp. 177–81; and B. B. Smuts and J. M. Watanabe, 'Social Relationships and Ritualized Greetings in Adult Male Baboons (*Papio cynocephalus anubis*)', *International Journal of Primatology*, 11:2 (1990), pp. 147–72.
22. V. Sommer, P. Schauer and D. Kyriazis, 'A Wild Mixture of Motivations: Same-Sex Mounting in Indian Langur Monkeys, in Sommer and Vasey (eds), *Homosexual Behaviour in Animals*, pp. 238–72.
23. The alliance formation hypothesis about same-sex sexuality certainly needs more supporting data. We strongly disagree, however, with Sommer's allegation that, as a hypothesis, it would be 'so ubiquitous as to be untestable' because its predictions are unclear. Curiously, many of the chapters in *Homosexual Behaviour in Animals* in fact support the alliance formation hypothesis.
24. Kirkpatrick, 'The Evolution of Human Homosexual Behavior'.
25. Murray, *Homosexualities*; Crompton, *Homosexuality and Civilization*; and Herdt (ed), *Ritualized Homosexuality in Melanesia*.
26. M. Rocke, *Forbidden Friendships: Homosexuality and Male Culture in Renaissance Florence* (Oxford: Oxford University Press, 1996).
27. For an overview and evaluation of these functions, see Vasey and Sommer, 'Homosexual Behaviour in Animals: Topics, Hypotheses and Research Trajectories'; and N. W. Bailey and M. Zuk, 'Same-sex Sexual Behaviour and Evolution', *Trends in Ecology and Evolution*, 24:8 (2009), pp. 439–46.
28. M. E. Perry, 'The "Nefarious Sin" in Early Modern Seville', in K. Gerard and G. Hekma (eds), *The Pursuit of Sodomy: Male Homosexuality in Renaissance and Enlightenment Europe* (London: Haworth Press, 1989), pp. 67–90.
29. L. Newson and P. J. Richerson, 'Why Do People Become Modern: A Darwinian Mechanism', *Population and Development Review*, 35:1 (2009), pp. 117–58.
30. Adriaens and De Block, 'The Evolution of a Social Construction'.
31. The contemporary evolutionary literature about homosexuality thrives on exactly this question. Here's a recent example: 'One especially puzzling fact regards the long-standing persistence of this apparently detrimental trait, with the associated stability of polymorphic human populations; this is a possible "Darwinian paradox": since male homosexuals don't mate with the opposite sex, shouldn't any "genes promoting homosexuality" have died out of the population by now?' (A. Camperio-Ciani, P. Cermelli and G. Zanzotto, 'Sexually Antagonistic Selection in Human Male Homosexuality', *PLoS ONE*, 3:6 (2008), e2282.
32. For an authoritative and critical overview of biological theories about same-sex sexuality, see W. Byne and B. Parsons, 'Sexual Orientation: The Biologic Theories Reappraised', *Archives of General Psychiatry*, 50 (1993), p. 228–39; and W. Byne, 'Why We Cannot Conclude Sexual Orientation Is a Biological Phenomenon' in J. K. Davidson and N. B. Moore (eds), *Speaking of Sexuality. Interdisciplinary Readings* (Oxford: Oxford University Press, 2007).
33. J. Henrich, S. J. Heine and A. Norenzayan, 'Most People Are Not WEIRD', *Nature*, 466 (2010), pp. 29.

34. Murray, *Homosexualities*.
35. R. Boyd and P. J. Richerson, *Culture and the Evolutionary Process* (Chicago, IL: University of Chicago Press, 1985).
36. Kirkpatrick, 'The Evolution of Human Homosexual Behavior', p. 397.
37. Boyd and Richerson, *Culture and the Evolutionary Process*; and P. J. Richerson and R. Boyd, *Not By Genes Alone: How Culture Transformed Human Evolution* (Chicago, IL: University of Chicago Press, 2005).
38. The *kin influence hypothesis* does not contradict the *alliance formation hypothesis* so it is best thought of as complimentary rather than as an alternative evolutionary explanation. For more information about the *kin influence hypothesis*, see L. Newson, T. Postmes, S. E. Lea and P. Webley, 'Why Are Modern Families Small? Toward an Evolutionary and Cultural Explanation for the Demographic Transition', *Personality and Social Psychology Review*, 9:4 (2005), pp. 360–75; and Newson and Richerson, 'Why Do People Become Modern: A Darwinian Mechanism'.
39. See for example Boyd and Richerson, *Culture and the Evolutionary Process*; and L. L. Cavalli-Sforza and M. W. Feldman, *Cultural Transmission and Evolution: A Quantitative Approach* (Princeton, NJ: Princeton University Press, 1981).
40. Richerson and Boyd, *Not By Genes Alone*; and L. Newson, P. J. Richerson and R. Boyd, 'Cultural Evolution and the Shaping of Cultural Diversity', in S. Kitayama and D. Cohen (eds), *The Handbook of Cultural Psychology* (New York: Guilford Press, 2007), pp. 454–76.
41. M. J. West-Eberhard, *Developmental Plasticity and Evolution* (Oxford: Oxford University Press, 2004).
42. G. Csibra and G. Gergely, 'Natural Pedagogy as Evolutionary Adaptation', *Philosophical Transactions of the Royal Society B*, 366 (2011), pp. 1149–57.
43. C. J. Lumsden and E. O. Wilson, *Genes, Mind, and Culture: The Coevolutionary Process* (Cambridge, MA: Harvard University Press, 1981).
44. R. Nisbett and S. Gurwitz, 'Weight, Sex, and the Eating Behavior of Human Newborns', *Journal of Comparative and Physiological Psychology*, 73 (1970), pp. 245–53.
45. M. J. Tovee, V. Swami, A. Furnham and R. Mangalparsad, 'Changing Perceptions of Attractiveness as Observers Are Exposed to a Different Culture', *Evolution and Human Behavior*, 27 (2006), pp. 443–56.
46. M. Apostolou, 'Sexual Selection under Parental Choice: The Role of Parents in the Evolution of Human Mating', *Evolution and Human Behavior*, 28 (2007), pp. 403–9; and M. Apostolou, 'Bridewealth as an Instrument of Male Parental Control over Mating: Evidence from the Standard Cross-Cultural Sample', *Journal of Evolutionary Psychology*, 8 (2010), pp. 205–16.
47. R. S. Walker, K.R. Hill, M.V. Flinn and R. M. Ellsworth, 'Evolutionary History of Hunter-Gatherer Marriage Practices', *PloS ONE*, 6:4 (2011), e19066.
48. M. Borgerhoff-Mulder, 'Behavioural Ecology in Traditional Societies', *Trends in Ecology and Evolution*, 3:10 (1988), pp. 260–4; L. Cronk, 'Human Behavioral Ecology', *Annual Review of Anthropology*, 20 (1991), pp. 25–53; B. S. Low, *Why Sex Matters* (Princeton, NJ: Princeton University Press, 2000); and E. Voland, 'Evolutionary Ecology of Human Reproduction', *Annual Review of Anthropology*, 27 (1998), pp. 347–74.
49. In north-western Europe during the seventeenth and eighteenth centuries, for example, marriage was often delayed until women were in their middle to late twenties and men in their thirties. Spinsters and bachelors were expected to help their parents or siblings and some remained unmarried. A combination of strong sanctions against unmarried women giving birth, strict monogamy and a belief that couples should not get married

until they had the means to set up an independent household served to regulate birth rates. See for example A. J. Coale, 'The Decline of Fertility in Europe since the 18th Century as a Chapter in Demographic History', in A. J. Coale and S. C. Watkins (eds), *Decline of Fertility in Europe* (Princeton, NJ: Princeton University Press, 1986), pp. 1–30; and J. Hajnal, 'Two Kinds of Pre-industrial Household Formation System', *Population and Development Review*, 8:3 (1982), pp. 449–94.

50. Newson and Richerson, 'Why do People become Modern: A Darwinian Mechanism'.
51. See for example J. Weeks, *The World we have Won: The Remaking of Erotic and Intimate Life* (Abingdon, Oxon: Routledge, 2007).
52. R. Inglehart and C. Welzel, *Modernization, Cultural Change, and Democracy: The Human Development Sequence* (New York: Cambridge University Press, 2005); and A. Inkeles and D. H. Smith, *Becoming Modern: Individual Change in Six Developing Countries* (Cambridge, MA: Harvard University Press, 1974).
53. K. Davis, 'Kingsley Davis on Reproductive Institutions and the Pressure for Population', *Population and Development Review*, 23:3 (1937/97), pp. 611–24.
54. S. C. Watkins, 'From Local to National Communities: The Transformation of Demographic Regimes in Western-Europe, 1870–1960', *Population and Development Review*, 16:2 (1990), pp. 241–72; and B. Wilson, 'Aspects of Secularization in the West', *Japanese Journal of Religious Studies*, 3:4 (1976), pp. 259–76.
55. B. Anderson, *Imagined Communities: Reflections on the Origin and Spread of Nationalism* (London: Verso, 1991).
56. W. D. Hamilton, 'Genetic Evolution of Social Behavior', *Journal of Theoretical Biology*, 7:1 (1964), pp. 1–52.
57. Kirkpatrick, 'The Evolution of Human Homosexual Behavior'.
58. Murray, *Homosexualities*.
59. P. Zachar, 'Psychiatric Disorders Are Not Natural Kinds', *Philosophy, Psychiatry & Psychology*, 7 (2000), pp. 167–82.
60. N. Haslam and S. R. Levy, 'Essentialist Beliefs about Homosexuality: Structure and Implications for Prejudice', *Personality and Social Psychology Bulletin*, 32:4 (2006), pp. 471–85.
61. Borris and Rousseau (eds), *The Sciences of Homosexuality in Early Modern Europe*.
62. Stein, *The Mismeasure of Desire*.
63. N. Haslam, L. Rothschild and D. Ernst, 'Are Essentialist Beliefs Associated With Prejudice?', *British Journal of Social Psychology*, 41 (2002), pp. 87–100.
64. J. P. De Cecco and D. A. Parker (eds), *Sex Cells and Same-Sex Desire: The Biology of Sexual Preference* (Binghamton, NY: Haworth Press, 1995). Nineteenth- and early twentieth-century activists like Ulrichs and Hirschfeld were literate and lived in big German cities. Consequently, it is likely that they would have had much more exposure to non-kin influence and earlier exposure to massive non-kin influence than most people of their time.
65. Byne, 'Sexual Orientation: The Biologic Theories Reappraised' and 'Why we Cannot Conclude Sexual Orientation is a Biological Phenomenon'.
66. A. Kinsey, W. Pomeroy and C. Martin, *Sexual Behavior in the Human Male* (Philadelphia, PA: W.B. Saunders Company, 1948), p. 639.
67. Hacking, *The Social Construction of What?*
68. R. Bayer, *Homosexuality and American Psychiatry: The Politics of Diagnosis* (Princeton, NJ: Princeton University Press, 1981).
69. Hacking, *The Social Construction of What?*

70. Oosterhuis, *Stepchildren of Nature*.
71. Essentializing homosexuality is intricately related, for example, to people's prejudices about it. Early twentieth-century gay activists conceptualized male love as a disease, rather than a vice or a sin, in order to secure a more humane treatment of homosexuals. Yet essentializing is often a double-edged sword. As Haslam and Levy note: '[A]ttributing deviant behaviour to uncontrollable causes such as genes is sometimes linked to tolerance, but biological explanations also have been shown to promote greater endorsement of gender and racial stereotypes, to gender and racial prejudice, and to more stigmatizing attitudes toward people'. See Haslam and Levy, 'Essentialist Beliefs about Homosexuality', p. 482.
72. S. Gelman, J. Coley and G. Gottfried, 'Essentialist Beliefs in Children: The Acquisition of Concepts and Theories', in L. Hirschfeld and S. Gelman (eds.), *Mapping the Mind: Domain Specificity in Cognition and Culture* (Cambridge: Cambridge University Press), pp. 341–66.
73. J. Dupré, *The Disorder of Things: Metaphysical Foundations of the Disunity of Science* (Harvard, MA: Harvard University Press, 1993).
74. H. Barrett, 'On the Functional Origins of Essentialism', *Mind and Society*, 3 (2001), pp. 1–30.
75. See for example N. Haslam, J. Rothschild and D. Ernst, 'Essentialist Beliefs About Social Categories', *British Journal of Social Psychology*, 39 (2000), pp. 113–27.
76. See for example N. Haslam and D. Ernst, 'Essentialist Beliefs About Mental Disorders', *Journal of Social and Clinical Psychology* 21 (2002), pp. 628–44; and M. Keller and G. Miller, 'Resolving the Paradox of Common, Harmful, Heritable Mental Disorders: Which Evolutionary Genetic Models Work Best?', *Behavioral and Brain Sciences*, 29 (2006), pp. 385–452.
77. F. Gil-White, 'Are Ethnic Groups Biological "Species" to the Human Brain? Essentialism in Our Cognition of Some Social Categories', *Current Anthropology*, 42 (2001), pp. 515–54.
78. D. Sperber, *Explaining Culture: A Naturalistic Approach* (London: Blackwell, 1996).
79. P. R. Adriaens and A. De Block, 'Why We Essentialize Mental Disorders', *Journal of Medicine & Philosophy* (forthcoming).
80. Gil-White, 'Are Ethnic Groups Biological "Species" to the Human Brain?'
81. R. Porter, *A Social History of Madness: The World through the Eyes of the Insane* (New York: E.P. Dutton, 1989).
82. American Psychiatric Association, *Diagnostic and Statistical Manual: Mental Disorders (First Edition)* (Washington: American Psychiatric Press, 1952); American Psychiatric Association, *Diagnostic and Statistical Manual of Mental Disorders (Second Edition)* (Washington: American Psychiatric Press, 1968).
83. Bayer, *Homosexuality and American Psychiatry*.
84. Oosterhuis, *Stepchildren of Nature*.
85. Foucault, *The History of Sexuality*, p. 43.
86. R. Bayer, 'Politics, Science, and the Problem of Psychiatric Nomenclature: a Case Study of the American Psychiatric Association referendum on homosexuality', in T. Engelhardt and A. Caplan (eds), *Scientific Controversies: Case-Studies in the Resolution and Closure in Science and Technology* (Cambridge: Cambridge University Press, 1987), pp. 381–400.

6 Munévar, 'Darwinism and Homosexuality'

1. See Chapter 8 of this volume.
2. B. Bagemihl, *Biological Exuberance: Animal Homosexuality and Natural Diversity* (New York: St Martin's Press, 1999).
3. W. D. Hamilton, 'The Genetical Evolution of Social Behaviour I and II', *Journal of Theoretical Biology*, 7 (1964), pp. 1–16, 17–52.
4. E. O. Wilson, *On Human Nature* (Cambridge: Cambridge University Press, 1978), 142–8.
5. Charles Darwin published the first edition of *The Origin of Species* in 1859 and the sixth edition in 1872. In this essay I will use the Mentor edition with the Introduction by Julian Huxley: Charles Darwin, *The Origin of Species* (New York: New American Library, 1958), pp. 250–6.
6. Ibid., p. 252.
7. Ibid., p. 251.
8. Ibid., p. 253.
9. Wilson, *Human Nature*, pp. 144–7.
10. S. J. Gould and R. C. Lewontin, 'The Spandrels of San Marco and the Panglossian Paradigm: A Critique of the Adaptationist Programme', *Proceedings of the Royal Society of London B*, 205:1161 (1979), pp. 581–98.
11. Even sympathetic readers of Wilson's use of kin selection in this context admit that his hypothesis is at best suggestive and that there is little evidence for it. See for example, M. Ruse, *Homosexuality* (Oxford: Basil Blackwell, 1988), pp. 130–49.
12. M. Ridley, *The Red Queen: Sex and the Evolution of Human Nature* (New York: HarperCollins, 2003), p. 72; first published by Penguin Books in 1993. See also G. E. Hutchinson, 'A Speculative Consideration of Certain Possible Forms of Sexual Selection in Man', *American Naturalist*, 93:869 (1959): pp. 81–91.
13. A. Camperio-Ciani, F. Corna and C. Capiluppi, 'Evidence for Maternally Inherited Factors Favouring Male Homosexuality and Promoting Female Fecundity', *Proceedings of the Royal Society of London, Series B: Biological Sciences*, 271 (2004), pp. 2217–21.
14. R. C. Pillard and J. D. Weinrich, 'Evidence of Familial Nature of Male Homosexuality', *Archives of General Psychiatry*, 43 (1986), pp. 808–12.
15. J. M. Bailey and D. S. Benishay, 'Familial Aggregation of Female Sexual Orientation', *American Journal of Psychiatry*, 150 (1993), pp. 272–7.
16. Simon LeVay's Website: 'The Biology of Sexual Orientation', which offers a summary of the latest research in the field and commentary by LeVay, updated April, 2009: www.simonlevay.com/the-biology-of-sexual-orientation: 8 [accessed 13 October 2009].
17. J. M. Bailey and R. C. Pillard, 'Genetics of Human Sexual Orientation', *Annual Review of Sex Research*, 6 (1995), pp. 126–50; and J. M. Bailey, R. C. Pillard, M. C. Neale and Agyei, 'Y Heritable Factors Influence Sexual Orientation in Women', *Archives of General Psychiatry* (1993), p. 50.
18. LeVay points out an additional problem: ascertainment bias, i.e. the likelihood that recruitment fliers that ask for homosexuals with twins may appeal more to those whose twins are also homosexual. 'Sexual Orientation', p. 9.
19. J. P. Macke, N. Hu, S. Hu, M. Bailey, V. L. King, T. Brown, D. Hamer and J. Nathans, 'Sequence Variation in the Androgen Receptor Gene is Not a Common Determinant of Male Sexual Orientation', *American Journal of Human Genetics*, 53 (1993): pp. 844–52.

20. D. H. Hamer, S. Hu, V. L. Magnuson, N. Hu and A. M Pattatucci, 'A Linkage Between DNA Markers on the X chromosome and Male Sexual Orientation', *Science*, 261 (1993), pp. 321–7.
21. G. Rice, C. Anderson, N. Risch and G. Ebers, 'Male Homosexuality: Absence of Linkage to Microsatellite Markers at Xq28', *Science*, 284 (1999), pp. 665–7. Also B. S. Mustanski, M. G. Dupree, C. M. Nievergelt, S. Bocklandt, N. J. Schork and D. H. Hamer, 'A Genomewide Scan of Male Sexual Orientation', *Human Genetics*, 116 (2005), pp. 272–8.
22. Darwin, *The Origin of Species*, p. 58.
23. Ibid., p. 132.
24. See, for example, a standard textbook such as L. H. Hartwell, L. Hood, M. L. Goldberg, A. E. Reynolds and L. M. Silver, *Genetics: From Genes to Genomes* (New York: McGraw-Hill, 2010), ch. 3.
25. LeVay, 'Sexual Orientation', p. 7.
26. The following account borrows heavily from Simon LeVay, *The Sexual Brain* (Cambridge, MA: MIT Press, 1993), pp. 17–29.
27. Estrogens and progestagens are the main classes of female sex steroids. Human female development at puberty is guided by hormones such as oestrogen. Subsequently, oestrogen and progesterone are of particular importance.
28. R. B. Simerly, 'Wired for Reproduction: Organization and Development of Sexually Dimorphic Circuits in the Mammalian Forebrain', *Annual Review of Neuroscience*, 25 (2002), pp. 507–36.
29. Simon LeVay, *Queer Science* (Cambridge, MA: MIT Press, 1996), p. 166.
30. L. G. Clemens, B. A. Gladue and L. P. Coniglio, 'Prenatal Endogenous Androgenic Influences on Masculine Sexual Behavior and Genital Morphology in Male and Female Rats', *Hormones and Behavior*, 10 (1978), pp. 40–53. Also: R. L. Meisel and I. L. Ward, 'Fetal Female Rats are Masculinized by Male Littermates Located Caudally in the Uterus', *Science*, 213 (1981), pp. 239–42. Such rats probably receive testosterone from the male directly upstream from them. For commentary see LeVay, *Sexual Brain*, pp. 89–90.
31. LeVay, *Sexual Brain*, p. 89.
32. Ibid., p. 90.
33. LeVay credits Richard Anderson for the discovery. R. H. Anderson, D. E. Fleming, R. W. Rhees and E. Kinghorn, 'Relationships between Sexual Activity, Plasma Testosterone, and the Volume of the Sexually Dimorphic Nucleus of the Preoptic Area in Prenatally Stressed and Non-Stressed Rats', *Brain Research*, 370 (1986), pp. 1–10.
34. For LeVay's detailed account consult *Sexual Brain* (particularly pp. 71–81) and *Queer Science* (particularly pp. 129–47). The report of the original study is S. LeVay, 'A Difference in Hypothalamic Structure between Heterosexual and Homosexual Men', *Science*, 253 (1991), pp. 1034–7.
35. The equivalent structure in females is the Ventromedial Nucleus, also in the hypothalamus.
36. J. A. Resko, A. Perkins, C. E. Roselli, J. A. Fitzgerald, J. V. A. Choate and F. Stormshak, 'Aromatase Activity and Androgen Receptor Content of Brains from Heterosexual and Homosexual Rams', *Proceedings of the Endocrine Society, 77th Annual Meeting* (June 1995), p. 135.
37. C. E. Roselli, K. Larkin, J. A. Resko, J. N. Stellflug and F. Stormshak, 'The Volume of a Sexually Dimorphic Nucleus in the Ovine Medial Preoptic Area/Anterior Hypothalamus Varies with Sexual Partner Preference', *Endocrinology*, 145 (2004), pp. 478–83.

38. See W. Byrne and B. Parsons, 'Human Sexual Orientation: The Biologic Theories Reappraised', *Archives of General Psychiatry*, 50 (March 1993), pp. 228–38.
39. W. Byne, S. Tobet, L. A. Mattiace, M. S. Lasco, E. Kemether, M. A. Edgar, S. Morgello, M. S. Buchsbaum, and L. B. Jones, 'The Interstitial Nuclei of the Human Anterior Hypothalamus: An Investigation of Variation with Sex, Sexual Orientation, and HIV Status', *Hormones and Behavior*, 40 (2001), pp. 86–92. My emphasis on both quotations. Now, the difference reported by Byne is not as large as that originally reported by LeVay. But then, both samples were small.
40. I. Savic and P. Lindstrom, 'PET and MRI Show Differences in Cerebral Asymmetry and Functional Connectivity between Homo- and Heterosexual Subjects,' *PNAS*, 105 (2008), pp. 403–8. For LeVay's commentary see 'Sexual Orientation', p. 25.
41. A. Perkins, J. A. Fitzgerald, and G. E. Moss, 'A Comparison of LH Secretion and Brain Stradiol Receptors in Heterosexual and Homosexual Rams and Female Sheep', *Hormones and Behavior*, 29 (1995), pp. 31–41.
42. C. L. Moore, 'Maternal Contribution to the Development of Masculine Sexual Behavior in Laboratory Rats', *Developmental Psychobiology*, 17 (1984), pp. 347–56.
43. LeVay, *Sexual Brain*. p. 92.
44. R. W. Goy, K. Wallen and D. A. Goldfoot, 'Social Factors Affecting the Development of Mounting Behavior in Male Rhesus Monkeys', in *Reproductive Behavior*, ed. W. Montagna and W. A. Sadler (New York: Plenum Press, 1974), pp. 223–47. LeVay's comments on this study, as well as his account of how the brain controls puberty, are to be found in *Sexual Brain*, pp. 93–5.
45. As reported in Bruce Bagemihl, *Biological Exuberance*, pp. 391–3.
46. B. M. Frans de Waal, 'Bonobo Sex and Society', *Scientific American* (March 1995), pp. 82–8. LeVay, *Queer Science*, pp. 206–9. LeVay's main source seems to be E. S. Savage-Rumbaugh and B. J. Wilkerson, 'Socio-Sexual Behavior in *Pan paniscus* and *Pan troglodytes*: A Comparative Study', *Journal of Human Evolution*, 7 (1978), pp. 327–44.
47. One might be tempted to treat the influence of social factors on the sexual behavior of giraffes, bonobos and other animals as examples of social construction of sexual preference. In the most common understanding of that term, i.e. a social process independent of biology, that conclusion would not be quite right, since the conclusions of the relevant studies are generalized to the species, and since their behavior is thus considered species wide, it should still be thought to be influenced by their respective genomes. Some degree of social construction in the case of humans, however, would be consistent with the remarks made in the text.
48. LeVay, *Queer Science*, p. 134.
49. Again, what we would find would be not that all men with such structural differences would be exclusive homosexuals, but that the probability of being so would be greater, given such differences.
50. A. Kinsey, W. Pomeroy and C. Martin, Sexual Behavior in the Human Male (Philadelphia, PA: W. B. Saunders, 1948); and A. Kinsey, W. Pomeroy, C. Martin and P. Gebhard, Sexual Behavior in the Human Female (Philadelphia, PA: W. B. Saunders, 1953).
51. The following studies re-analysed Kinsey's work: P. H. Gebhard, 'Incidence of Overt Homosexuality in the United States and Western Europe', in *National Institute of Mental Health Task Force on Homosexuality: Final Report and Background Papers*, ed. J. M. Livingood (Rockville, MD: National Institute of Mental Health, 1972); J. Gagnon and W. Simon, *Sexual Conduct: The Social Sources of Human Sexuality* (Chicago, IL: Aldine, 1973); P. H. Gebhard and A. B. Johnson, *The Kinsey Data: Marginal Tabulations of*

1938–1963 Interviews Conducted by the Institute for Sex Research (Philadelphia, PA: W.B. Saunders,1979); D. McWhirter, S. Sanders and J. Reinisch (eds), *Homosexuality/ Heterosexuality*, The Kinsey Institute Series (New York: Oxford University Press, 1990). Some of the more recent studies include the following: S. Janus and C. Janus, *The Janus Report on Sexual Behavior* (New York: John Wiley and Sons, 1993); J. Billy, K. Tanfer, W. Grady and D. Klepinger, 'The Sexual Behavior of Men in the United States', *Family Planning Perspectives*, 25:2 (1993), pp. 52–60; E. Laumann, J. H .Gagnon, R. T. Michael and S. Michaels, *The Social Organization of Sexuality: Sexual Practices in the United States* (Chicago, IL: University of Chicago Press, 1994); R. L. Sell, J. A. Wells and D. Wypij, 'The Prevalence of Homosexual Behavior and Attraction in the United States, the United Kingdom and France: Results of National Population-Based Samples', *Archives of Sexual Behavior*, 24:3 (1995), pp. 235–48; M. Diamond, 'Homosexuality and Bisexuality in Different Populations', *Archives of Sexual Behavior*, 22:4 (1993), pp. 291–310.

52. Those are the kinds or co-regulations or hierarchical genetic relationships described by *pleiotropy* and *epistasis*.
53. LeVay anticipated in part the thrust of my argument when he speculated in *Queer Science* that 'there is a broad molecular diversity within the pathways of sexual development that permits far more subtle variations in a person's sexuality than simply "all male" or "all female"' (p. 279). Indeed we can find variation in the tertiary and quaternary folding of proteins already, and thus a diversity of chemical action at the molecular level. As Gunther Stent pointed out in 1981, 'the subsequent folding of the completed polypeptide chain into its specific tertiary structure lacks programmatic character, since the three dimensional conformation of the molecule is the automatic consequence of its *contextual situation* and has no isomorphic correspondent in the DNA' (my italics) – 'Strength and Weakness of the Genetic Approach to the Development of the Nervous System', *Annual Review of Neuroscience*, 4 (1981) pp. 163–94, on p. 187.
54. The fertility rate of exclusive homosexuality need not be zero, even if at first that appears to be a contradiction in terms otherwise. One reason is that homosexuals may still perform as members of heterosexual couples. Lesbians may for social reasons be expected to bear children. And male homosexuals may donate their sperm or engage in copulation with their wives as if they were donating sperm. Besides, those genes will also interact with the environment and with developmental noise, and thus they may also be expressed in a distribution of traits, even if most carriers do behave as exclusive homosexuals.
55. Darwin, *The Origin of Species*, p. 132.

7 Lloyd, 'The Evolution of Female Orgasm: New Evidence and Response to Feminist Critiques'

1. E. A. Lloyd, 'The Nature of Darwin's Support for the Theory of Natural Selection', *Philosophy of Science*, 50 (1983), pp. 112–29.
2. A. Kinsey, et al., *Sexual Behavior in the Human Female* (Philadelphia, PA: W. B. Saunders, 1953).
3. K. Dawood et al., 'Genetic and Environmental Influences on the Frequency of Orgasm in Women', *Twin Research and Human Genetics*, 8:1 (2005), pp. 27–33; K. M. Dunn, L. F. Cherkas and T. D. Spector, 'Genetic Influences on Variation in Female Orgasmic Function: A Twin Study', *Biology Letters*, 1:3 (2005), pp. 260–3; E. A. Lloyd, *The Case*

of the Female Orgasm: Bias in the Science of Evolution (Cambridge, MA: Harvard University Press, 2005).

4. B. P. Zietsch et al., 'Female Orgasm Rates are Largely Independent of Other Traits: Implications for "Female Orgasmic Disorder" and Evolutionary Theories of Orgasm', *Journal of Sexual Medicine*, 8:8 (2011), pp. 2305–16.
5. Lloyd, *The Case of the Female Orgasm*.
6. E. A. Lloyd, 'Pre-theoretical Assumptions in Evolutionary Explanations of Female Sexuality', *Philosophical Studies*, 59 (January 1993), pp. 139–53.
7. D. Morris, *The Naked Ape: A Zoologist's Study of the Human Animal* (Canada, Bantam Books, 1967). p. 65.
8. See Lloyd, *The Case of the Female Orgasm: Bias in the Science of Evolution*.
9. Figure from Dawood et al., 'Genetic and Environmental Influences on the Frequency of Orgasm in Women', p. 29.
10. Kinsey, *Sexual Behavior in the Human Female*; Zietsch, 'Female Orgasm Rates are Largely Independent of Other Traits'.
11. E. A. Lloyd, 'Confirmation of Evolutionary and Ecological Models', *Biology and Philosophy*, 2:3 (1987), pp. 277–93.
12. H. E. Longino, *Science as Social Knowledge: Values and Objectivity in Scientific Inquiry* (Princeton, NJ: Princeton University Press, 1990). p. 129.
13. S. B. Hrdy, *The Woman that Never Evolved* (Cambridge, MA: Harvard University Press, 1981).
14. Lloyd, *The Case of the Female Orgasm*, pp. 95–105.
15. M. Kosfeld et al., 'Oxytocin Increases Trust in Humans', *Nature*, 435 (2005), pp. 673–6; W. Blaicher et al., 'The Role of Oxytocin in relation to Female Sexual Arousal', *Gynecologic and Obstetric Investigation*, 47 (1999), pp. 125–6; L. Wildt et al., 'Sperm Transport in the Human Female Genital Tract and its Modulation by Oxytocin as Assessed by Hysterosalpingoscintigraphy, Hysterotonography, Electrohysterography and Doppler Sonography', *Human Reproduction Update*, 4:5 (1998), pp. 655–66; B. R. Komisaruk et al., *The Science of Orgasm* (Baltimore, MD: Johns Hopkins University Press, 2006).
16. Wildt et al., 'Sperm Transport in the Human Female Genital Tract'.
17. Blaicher et al., 'The Role of Oxytocin in relation to Female Sexual Arousal'; Komisaruk et al., *The Science of Orgasm*.
18. Levin, 'Can the Controversy about the Putative Role of the Human Female Orgasm in Sperm Transport be Settled with our Current Physiological Knowledge of Coitus?', *Journal of Sexual Medicine*, 8:6 (2011), pp. 1566–78, on p. 1573.
19. J. Alcock, 'Unpunctuated Equilibrium in the Natural History Essays of Stephen Jay Gould', *Evolution and Human Behavior*, 19 (1998), pp. 321–36; P. Sherman, 'The Clitoris Debate and the Levels of Analysis', *Animal Behaviour*, 37:4 (1989), pp. 697–8; R. R. Baker and M. A. Bellis, 'Human Sperm Competition: Ejaculate Manipulation by Females and a Function for the Female Orgasm', *Animal Behaviour*, 46 (1993), pp. 887–909; R. Thornhill et al., 'Human Female Orgasm and Mate Fluctuating Asymmetry', *Animal Behaviour*, 50 (1995), pp. 1601–15; D. A. Puts and K. Dawood, 'The Evolution of Female Orgasm: Adaptation or Byproduct?', *Twin Studies and Genetic Research*, 9 (2006), pp. 467–72; S. Linquist, 'Sometimes an Orgasm is just an Orgasm', *Metascience*, 15 (2006), pp. 411–9; D. Barash and J. E. Lipton, *How Women got their Curves and Other Just-So Stories* (New York: Columbia University Press, 2009).
20. S. W. Gangestad and G. J. Scheyd, 'The Evolution of Human Physical Attractiveness', *Annual Review of Anthropology*, 34 (2005), pp. 523–48.
21. Baker and Bellis, 'Human Sperm Competition'.

22. Levin, 'Can the Controversy about the Putative Role of the Human Female Orgasm in sperm Transport be Settled with our Current Physiological Knowledge of Coitus?' p. 1576.
23. Baker and Bellis, 'Human Sperm Competition'.
24. See Lloyd, *The Case of the Female Orgasm: Bias in the Science of Evolution*, pp. 198–209 for detailed discussion of their methods and data.
25. Ibid., pp. 209–16.
26. Wallen and I also found earlier that the length of the clitoris, a potentially important correlate of ease of orgasm with intercourse, was three times more variable than its homologue, the length of the penis. This is strong evidence that this aspect of the structure of the female genitals is not under direct selection, which will generally reduce variability in the trait.
27. See K. Wallen and E. A. Lloyd, 'Inappropriate Comparisons and the Weakness of Cryptic Choice: A Reply to Vincent J. Lynch and D. J. Hosken', *Evolution and Development*, 10:4 (2008), pp. 398–9. A. Pomiankowski and A. P. Moller, 'A Resolution to the Lek Paradox', *Proceedings of the Royal Society London B*, 260 (1995), pp. 21–9.
28. S. Bright, 'Nancy Drew and the Case of the Female Orgasm' (2005), at susiebright.blogs.com/susie_brights_journal_/2005/06/the_latest_edit.html [accessed 5 July 2011].
29. Kinsey, *Sexual Behavior in the Human Female*.
30. S. Fisher, *The Female Orgasm: Psychology, Physiology, Fantasy* (New York: Basic Books, 1973). Also, Wallen and Lloyd, unpublished data.
31. D. A. Goldfoot et al., 'Behavioral and Physiological Evidence of Sexual Climax in the Female Stump-Tailed Macaque (*Macaca arctoides*)', *Science*, 208 (1980), pp. 1477–8.
32. S. Chevalier-Skolnikoff, 'Homosexual Behavior in a Laboratory Group of Stumptail Monkeys: Forms, Contexts, and Possible Social Functions', *Archives of Sexual Behavior*, 5:6 (1976), pp. 511–27.
33. D. Symons, The Evolution of Human Sexuality (New York: Oxford University Press, 1979). pp. 88; Lloyd, *The Case of the Female Orgasm: Bias in the Science of Evolution*, pp. 101, 131.
34. Zietsch, 'Female Orgasm Rates are Largely Independent of Other Traits'.
35. S. J. Gould and R. C. Lewontin, 'The Spandrels of San Marco and the Panglossian Paradigm', *Proceedings of the Royal Society of London B*, 205 (1979), pp. 587–98.
36. P. Godfrey-Smith, 'Three Kinds of Adaptationism', in S. H. Orzack and E. Sober (eds.), *Adaptationism and Optimality* (Cambridge: Cambridge University Press, 2001), pp. 335–57, on p. 342
37. Alcock, 'Unpunctuated Equilibrium in the Natural History Essays of Stephen Jay Gould', p. 330. My emphasis, his emphasis.
38. D. Barash and J. E. Lipton, *How Women got their Curves and Other Just-So Stories*, p. 133.
39. Lloyd, *The Case of the Female Orgasm: Bias in the Science of Evolution*, p. 19.
40. Alcock (1998), Paul Sherman (1989), and David Barash and Judith Lipton (2009); Sarah Blaffer Hrdy (2005).
41. In *The Evolution of Desire: Strategies of Human Mating* (New York: Basic Books, 1994) (see discussion on pp. 75–6).
42. Fausto-Sterling et al., footnote 3, 1997, p. 416.
43. A. Fausto-Sterling, 'The Case of the Female Orgasm: Bias in the Science of Evolution (review)', *Journal of the History of the Behavioral Sciences* (2006), Doi: 10 1002/jhbs: 406-407. M. Zuk, 'Essay Review: The Case of the Female Orgasm: Bias in the Science of Evolution', *Perspectives in Biology and Medicine*, 49:2 (2006), pp. 294–8.
44. L. Gannett, 'The Case of the Female Orgasm: Bias in the Science of Evolution (review)', *Canadian Journal of Philosophy*, 37:4 (2007), pp. 619–38, on p. 625.

45. Gannett writes that 'epistemic space remains for feminist biologists to continue to explore adaptive explanations that treat female sexuality as autonomous and permit female sexual agency, even if such preferences – and accompanying antipathy toward Symons' service-with-a-smile account of female orgasm – are inspired in part by feminist politics' ('The Case of the Female Orgasm', p. 633). Earlier, she equated seeking an 'autonomous' explanation for female orgasm with 'seeking an evolutionary explanation of female orgasm that takes as its starting point female rather than male sexuality' (p. 623).
46. Ibid., p. 625.
47. Ibid.
48. 'They might be considered logically independent in the sense that the proposition "Women are motivated to have intercourse with men because they want to provide a service" ("Fries with that? Have a nice day!") could be false while the proposition "Female orgasm arose as the developmental outcome of selection pressure for male orgasm' could be *true*", (Gannett, 'The Case of the Female Orgasm', p. 625).
49. Ibid.
50. With the exception of Zietsch and Santtila 2011. Because Zietsch and Santtila used measures of male and female orgasm that were fundamentally different, and not expected to correlate – i.e., time to orgasm in men, compared to ability to orgasm at all in women – the data in this paper are not relevant to the case.
51. Dawood et al., 'Genetic and Environmental Influences on the Frequency of Orgasm in Women'; Dunn et al., 'Genetic Influences on Variation in Female Orgasmic Function: A Twin Study'.
52. Wallen and Lloyd, 'Inappropriate Comparisons and the Weakness of Cryptic Choice', a, b; K. Wallen and E. A. Lloyd, 'Female Sexual Arousal: Genital Anatomy and Orgasm in Intercourse', *Hormones and Behaviour*, 59 (2011), pp. 780–92; Levin, 'Can the Controversy about the Putative Role of the Human Female Orgasm in Sperm Transport be Settled with our Current Physiological Knowledge of Coitus?'; Zietsch et al., 'Female Orgasm Rates are Largely Independent of Other Traits'.
53. Lloyd, *The Case of the Female Orgasm*; E. A. Lloyd, 'Adaptationism in Action', delivered to the Philosophy Department, UC Santa Barbara, February 2011.

8 Burges, Cela-Conde and Nadal, 'Altruism and Sexual Selection'

1. C. Darwin, *The Descent of Man and Selection in relation to Sex* (London: John Murray, 1871).
2. E. O. Wilson, *Sociobiology: The New Synthesis* (Cambridge MA: Harvard University Press, 1975).
3. V. C. Wynne-Edwards, *Animal Dispersion in relation to Social Behaviour* (Edinburgh: Oliver & Boyd, 1962).
4. G. C. Williams, *Adaptation and Natural Selection* (Princeton NJ: Princeton University Press, 1966).
5. Wilson, *Sociobiology*; R. Dawkins, *The Selfish Gene* (Oxford: Oxford University Press, 1976.
6. K. N. Laland and G. R. Brown, *Sense and Nonsense* (Oxford, MA: Oxford University Press, 2002).
7. W. D. Hamilton, 'The Evolution of Altruistic Behaviour', *American Naturalist,* 97 (1963), pp. 354–6.
8. R. L. Trivers, 'The Evolution of Reciprocal Altruism', *Quarterly Review of Biology,* 46 (1971), pp. 35–57.

9. E. Fehr and S. Gächter, 'Altruistic Punishment in Humans', *Nature*, 415 (2002), pp. 137–40. J. H. Fowler, 'Altruistic Punishment and the Origin of Cooperation', *Proceedings of the National Academy of Sciences of the USA*, 102 (2005), pp. 7047–9.
10. Wynne-Edwards, *Animal Dispersion in relation to Social Behaviour*.
11. B. C. R. Bertram, 'Problems with Altruism', in K. S. C. S. Group (ed.), *Current Problems in Sociobiology* (Cambridge MA: Cambridge University Press, 1982), pp. 251–67. B. Voorzanger, 'Altruism in Sociobiology: A Conceptual Analysis', *Journal of Human Evolution*, 13 (1984), pp. 33–9. D. S. Wilson, 'On the Relationship between Evolutionary and Psychological Definitions of Altruism and Selfishness', *Biology and Philosophy*, 7 (1992), pp. 61–8. T. Settle, '"Fitness" and "Altruism": Traps for the Unwary, Bystander and Biologist Alike', *Biology and Philosophy*, 8 (1993), pp. 61–83.
12. Voorzanger, 'Altruism in Sociobiology: A Conceptual Analysis'; Wilson, 'On the Relationship between Evolutionary and Psychological Definitions of Altruism and Selfishness'; Settle, '"Fitness" and "Altruism"'.
13. J. E. Duffy, 'Eusociality in a Coral-Reef Shrimp', *Nature*, 381 (1996), pp. 512–14.
14. M. J. O'Riain, J. U. M. Jarvis and C. G. Faulkes, 'A Dispersive Morph in the Naked Mole-Rat', *Nature*, 380 (1996), pp. 619–21.
15. Dawkins, *The Selfish Gene*.
16. E. Sober and D. S. Wilson, *Unto Others: The Evolution and Psychology of Unselfish Behaviour* (Cambridge MA: Harvard University Press, 1998).
17. J. Maynard Smith, 'Evolution and the Theory of Games', *American Scientist*, 64 (1976), pp. 41–55.
18. Sober and Wilson, *Unto Others*.
19. Darwin, *The Descent of Man and Selection in relation to Sex*, ch. 4.
20. M. D. Hauser, 'Is Morality Natural?', *Newsweek*, http://www.news-week.com/id/158760 (2008).
21. F. J. Ayala, 'What the Biological Sciences Can and Cannot Contribute to Ethics', in F. J. Ayala and R. Arp (eds), *Contemporary Debates in Philosophy of Biology* (Oxford: Wiley-Blackwell, 2009).
22. Darwin, *The Descent of Man and Selection in relation to Sex*.
23. M. Ruse, 'The Biological Sciences Can Act as a Ground for Ethics', in F. J. Ayala and R. Arp (eds), *Contemporary Debates in Philosophy of Biology* (Oxford: Wiley-Blackwell, 2009).
24. Ibid.
25. F. J. Ayala, 'What the Biological Sciences Can and Cannot Contribute to Ethics', in Ayala and Arp (eds), *Contemporary Debates in Philosophy of Biology*.
26. Darwin, *The Descent of Man and Selection in relation to Sex*, p. 638.
27. Ibid., p. 638.
28. Ibid., p. 641.
29. A. Zahavi, 'Reliability in Communication Systems and the Evolution of Altruism', in B. Stonehouse and C. Perrins (eds), *Evolutionary Ecology* (London: MacMillan Press, 1977), pp. 253–60.
30. A. Paul, 'Sexual Selection and Mate Choice', *International Journal of Primatology*, 23 (2002), pp. 877–904.
31. I. Tessman, 'Human Altruism as a Courtship Display', *Oikos*, 74 (1995), pp. 157–8. G. F. Miller, 'Sexual Selection for Moral Virtues', *Quarterly Review of Biology*, 82 (2007), pp. 97–126.

32. Zahavi, 'Reliability in Communication Systems and the Evolution of Altruism'; Tessman, 'Human Altruism as a Courtship Display'; Miller, 'Sexual Selection for Moral Virtues'.
33. Miller, 'Sexual Selection for Moral Virtues'.
34. Ibid.
35. Paul, 'Sexual Selection and Mate Choice'.
36. H. Kokko, M. D. Jennions and R. Brooks, 'Unifying and Testing Models of Sexual Selection', *Annual Review of Ecology and Evolutionary Systems*, 37 (2006), pp. 43–66.
37. Ayala, 'What the Biological Sciences Can and Cannot Contribute to Ethics'; Miller, 'Sexual Selection for Moral Virtues'.
38. Miller, 'Sexual Selection for Moral Virtues'; Tessman, 'Human Altruism as a Courtship Display'.
39. Kokko et al., 'Unifying and Testing Models of Sexual Selection'.
40. Paul, 'Sexual Selection and Mate Choice'.
41. Miller, 'Sexual Selection for Moral Virtues'.
42. O. P. John and S. Srivastava, 'The Big-Five Trait Taxonomy: History, Measurement and Theoretical Perspectives', in L. Pervin and O. P. John (eds), *Handbook of Personality: Theory and Research*, 2nd edn (Guilford, 1999), pp. 102–38.
43. Ibid.
44. C. S. Lai, S. E. Fisher, J. A. Hurst, F. Vargha-Khadem and A. P. Monaco, 'A Forkhead-Domain Gene is Mutated in a Severe Speech and Language Disorder', *Nature*, 413 (2001), pp. 519–23.
45. W. Shu, H. Yang, L. Zhang, M. M. Lu and E. E. Morrisey, 'Characterization of a New Subfamily of Winged-Helix/Forkhead (Fox) Genes that are Expressed in the Lung and Act as Transcriptional Repressors', *Journal of Biological Chemistry*, 276 (2001), pp. 488–97.
46. K. L. Jang, W. J. Livesley and P. A. Vernon, 'Heritability of the Big Five Personality Dimensions and their Facets: A Twin Study', *Journal of Personality*, 64 (1996), pp. 577–91.
47. Ibid.
48. J. C. Flack and F. B. M. de Waal, '"Any Animal Whatever": Darwinian Building Blocks of Morality in Monkeys and Apes', *Journal of Consciousness Studies*, 7 (2000), pp. 1–29.

9 Bernal, 'The Role of Sex and Reproduction in the Evolution of Morality and the Law'

1. R. Wright, Nadie pierde: la teoría de los juegos y la lógica del destino humano (Barcelona: Tusquets, 2005).
2. E. O. Wilson, *Sobre la naturaleza humana* (México: Fondo de Cultura Económica, 1991).
3. For example, when they break palm leaves they need a lot of strength and only adult males have it, so they share them with the females and offspring; when they attack coatis' nests (who can double the size of capuchins monkeys), the masters let the others collect leftovers or take pieces and they even give some to beggars. F. De Waal, *Bien natural* (Barcelona: Herder, 1997).
4. Ibid.
5. De Waal tells about a high ranked female called Puist who took a risk when she helped her friend Luit to frighten away his rival Nikkie. But every time Nikkie had an important

conflict, he had the habit of harrassing his rivals' allies with the purpose of punishing them. This time, Nikkie harrassed Puist after the challenge; then Puist turned to Luit and extended her hand as if asking for help. Nevertheless, Luit did nothing to protect her. As soon as Nikkie left, Puist approached Luit howling in anger, ran after him and she even hit him.

6. The decrease of jungle habitats produced restrictions of food resources for leaf and fruit-eating primate species and as a consequence an increase in competition, while omnivorous had a selective advantage with the change of climate and environments; they simply changed their field of action to get the necessary food and/or increased the energetic products in their diets such as meat or fish and mollusks. A. Fontdevila, and A. Moya, *Evolución* (Madrid: Síntesis, 2003).
7. Forests and trees offer fruits, insects, eggs and honey; bush savannas have legumes and small animals for hunting such as young pigs, antelopes and hares; the underground of forests and savannas offer roots and tubers. Searching certain food becomes a competition with other monkeys, birds, bats while other foods such as nuts with hard cover or parts deep in the ground have fewer competitors. M. Brunet and P. Picq, 'La gran expansión de los australopitecos', in Y. Coppens (dir), *Los orígenes de la humanidad* (Madrid: Espasa, 2004), pp. 200–63.
8. Reichholf, thinks that our ancestors shared and competed with other carnivore animals for recent killed prey. Using stone chips they were able to open the skin and bones before they got rotten and were able to hide the pieces from other scavengers.
9. The findings about fauna indicate that our ancestors shared their habitat with other carnivores such as stripped and spotted hyenas and large felines such as lions, panthers, wild cats, giant cheetahs and saber-toothed tigers. Among herbivorous there are four genus of hogs, three of elephants, two of giraffes, antelopes, gazelles and buffaloes. Primates are also represented by relatives of the baboons, theropithecus and other species similar to today's colobus. M. Brunet and P. Picq, 'La gran expansión de los australopitecos'.
10. The relation between cooperation for hunting, prey robbing, prey transportation and chopping and the evolution of sharing behavior is due to the characteristics of the food: it is concentrated in one unit, it tends to decompose, it is too large to be consumed by one individual, its finding is unpredictable and its hunting requires certain strengths and abilities that depend on others to be successful. As a consequence collaboration is needed to get the food and this behaviour persists if there is sharing of the loot. F. De Waal, *Bien natural*.
11. Fontdevila and Moya, say that what seems to have happened in human evolution is that we *took more time to grow up*, not that the process is much slower; that is, it is a peramorphic process and not a paedomorphic one. Human growth is not slower than those of gorillas and chimpanzees, but it takes much more time and this leads to a hypermorphic structure with a delay at the final moment of the development for many structures or characteristics, 'an adult human is characterized by a larger brain, a postponed sexual maturity, a longer life expectancy and a larger body size than his/her ancestors. Nevertheless, some characteristics, such as body hair seem to be neotenous, which confirms that human evolution has been in mosaic.'
12. M. Domínguez, *El primate excepcional* (Barcelona: Ariel, 2002).
13. An Evolutionary Stable Strategy is a strategy that, when present in the majority of the individuals conform a population, cannot be replaced by another different alternative.
14. N. Acarin, *El cerebro del rey. Una introducción apasionante a la conducta humana* (Barcelona: RBA, 2001).

15. 'Almost in all 100 or more societies (hunter-gatherer) studied around the world men are responsible for most of the hunting and women are in charge of most of the vegetable gathering. Men form organized moving groups which go away from the settlements searching large pieces. Women participate in the capture of small animals and gather vegetables. Although men bring home high level proteins, women generally contribute most calories'. E. O. Wilson, *Sobre la naturaleza humana*.
16. 'Male gorillas have harems and they use their canines to fight for the females or to establish dominance'. D. C. Johanson, 'Cara a cara con la familia de Lucy', en AAVV, *Los orígenes del hombre* (México, D.F: Océano de México: National Geographic, 2003), pp. 24–45.
17. An important factor in *Homo erectus* is that the proportion in sexual dimorphism decreased considerably concerning body size. Males are just 20 to 30 per cent larger than females, which according to Lewin, would mean a significant reduction of competition between males, and this could be due to an increase of meat in the diet, more hunting, more cooperation or a direct selection of females.
18. Hostility between males is present when the females are in estrus. 'The elimination of estrus in primitive beings reduced the possibility of this competition and reinforced the alliances between male hunters'. E. O. Wilson, *Sobre la naturaleza humana*.
19. Frequent intercourse with just one male gives him reassurance that the offspring are his, and it can also be a strategy to reduce the probability of success in searching other mates; that is, it prevents the male from copulating outside the couple. D. P. Barash, and J. E. Lipton, *El Mito de la monogamia* (Madrid: Siglo XXI, 2003).
20. D. C. Johanson and E. Maitland, *El primer antepasado del hombre* (Bogotá: Planeta, 1981).
21. 'It is probable that the affective mechanisms which take place in the formation of the couple have evolved from adaptive forms present in our species ... In our species, the appropriation of the partner, the protection of offspring and the guardian role of the male may have contributed to the evolution of the couple and the mechanisms that supports it'. J. van H. Aram, 'Vivir en grupo', in Y. Coppens (dir), *Los orígenes de la humanidad* (Madrid: Espasa, 2004), pp. 198–239.
22. The size of human testis, five times larger than those of the gorilla and three times smaller than those of the chimpanzees, could indicate that there was a sperm competition between males, which is compatible with a monogamous species that shows a 'certain degree of female and male infidelity'. M. Ridley, *Evolution* (USA: Blackwell Publishing, 2004).
23. Sober and Sloan claim that all the theories explaining the evolution of altruism are based on the same process: a group of individuals that help others would have more descendants than those of a group whose members do not help others, even though this decreases the individual relative aptitude within the group.
24. In the *Paez* or *Nasa* indigenous community, individuals have interiorized reciprocity and solidarity behaviors to the point that, if they do not follow them, the consequence is that the individual feels sick. To recover the broken balance he/she must perform some tasks for the community that imply generosity and solidarity, while at the same time achieving spiritual healing. H. Gómez, *De la Justicia y el Poder Indígena* (Cauca, Colombia: Universidad del Cauca, 2000).
25. Gintis et al., 'Explaning Altruistic Behavior in Humans', *Evolution and Human Behavior*, 24 (2003), pp. 153–72.
26. M. A. Nowak and K. Sigmund, 'Evolution of Indirect Reciprocity', *Nature*, 437 (October 2005), pp. 1291–8.

27. In studies developed in fifteen small societies in four different continents in which the 'ultimatum' game was played, 'regardless of the cultural variations, the result was always far from what the rational analysis of selfish players predicted. Most people everywhere valued equitable results; it strongly contrasts with what selfish maximizes would do' and they conclude: 'evolution may have produced in human beings the emotions that affect these behaviors during the millions of years that human beings have been living in small groups. These emotions led us to behave in a way that in the long term, ended up being beneficial for us or for our group' Sigmund et al., 'La economía del juego limpio', *Investigación y Ciencia* (Marzo, 2002), pp. 22–7.
28. The development of jealousy would be consistent with the selection of monogamy, and with the analysis of potential losses in terms of biological efficiency, since a man would have a lot to lose by devoting his parental investment on the children of other men while 'sacrificing' greater biological efficiency by having children with other females; a female loses if her partner invests too little in her offspring because he also has to invest in the offspring of other females. R. A. Maier, *Comportamiento animal* (Madrid: Mc Graw Hill, 2001).
29. 'A classical anthropological study determined that only 4 out of 849 human societies did not show any interest for that surveillance through which men control their partners. In some societies, husbands even time the absences of their wives' D. P. Barash and J. E. Lipton, *El Mito de la monogamia*.
30. Based on interactive computer simulation models of 'Prisoner's Dilemma' games involving many people, evolutionary of *n*-people, Axelrod found that norms of behaviour emerge as the behavior results in greater adaptation: 'There is a norm in a given social situation as long as the individuals usually act in a given way and they are frequently punished when they do not act that way'.
31. In anthropology, a horde is a relatively small social group that consists of several families.
32. J. S. Bernal, *Evolución biológica de la moral y el derecho.* (Barranquilla- Bogotá: Editorial Universidad del Norte y Grupo editorial Ibañez, 2011).

10 Brown, 'Symmetry and Evolution: A Genomic Antagonism Approach'

1. C. R. Darwin, *The Variation of Animals and Plants under Domestication, Volume II*, 1st edn (London: John Murray , 1868), p. 322.
2. L. Van Valen, 'A Study of Fluctuating Asymmetry', *Evolution*, 16 (1962), pp. 125–42.
3. R. Thornhill and A. P. Møller, 'Developmental Stability, Disease and Medicine', *Biological Reviews*, 72 (1997), pp. 497–548; J. W. Chapman and D. Goulson, 'Environmental versus genetic influences on fluctuating asymmetry in the house fly, *Musca domestica*', *Biological Journal of the Linnean Society, 70* (2000), pp. 403–13; A. P. Møller, G. S. Sanotra and K. S. Vestergaard, 'Developmental Instability and Light Regime in Chickens (*Gallus gallus*)', *Applied Animal Behaviour Science, 62* (1999), pp. 57–71; J. T. Manning, 'Fluctuating Asymmetry and Body Weight in Men and Women: Implications for Sexual Selection', *Ethology and Sociobiology*, 16 (1995), pp. 145–53; L. W. Simmons, J. L. Tomkins and J. T. Manning, 'Sampling Bias and Fluctuating Asymmetry', *Animal Behaviour*, 49 (1995), pp. 1697–9.
4. P. A. Parsons, 'Fluctuating Asymmetry: A Biological Monitor of Environmental and Genomic Stress', *Heredity*, 68 (1992), pp. 361–4.

5. T. A. Markow, 'Evolutionary Ecology and Developmental Instability', *Annual Review of Entomology*, 40 (1995), pp. 105–20.
6. R. Thornhill and S.W. Gangestad, 'Human Fluctuating Asymmetry and Sexual Behavior', *Psychological Science*, 5 (1994), pp. 297–302; S. Van Dongen, R. Cornille and L. Lens, 'Sex and Asymmetry in Humans: What is the Role of Developmental Instability?', *Journal of Evolutionary Biology, 22* (2009), pp. 612–22.
7. J. W. Chapman and D. Goulson, 'Environmental versus Genetic Influences on Fluctuating Asymmetry in the House Fly, *Musca domestica*', *Biological Journal of the Linnean Society*, 70 (2000), 403–13; R. Thornhill and A. P. Møller, 'Developmental Stability, Disease and Medicine', *Biological Reviews*, 72 (1997), pp. 497–548; J. T. Manning, K. Koukourakis and D. A. Brodie, 'Fluctuating Asymmetry, Metabolic Rate and Sexual Selection in Human Males'. *Evolution and Human Behaviour*, 18 (1997), pp. 15–21; J. T. Manning and L. J. Pickup, 'Symmetry and Performance in Middle Distance Runners', *International Journal of Sports Medicine*, 19 (1998), pp. 205–9; S. M. Hughes, M. A. Harrison and G. G. Gallup, 'The Sound of Symmetry: Voice as a Marker of Developmental Instability', *Evolution and Human Behaviour*, 23 (2002), pp. 173–80; S. W. Gangestad and R. Thornhill, 'The Evolutionary Psychology of Extrapair Sex: The Role of Fluctuating Asymmetry', *Evolution of Human Behaviour*, 18 (1997), pp. 69–88.
8. S. W. Gangestad and R. Thornhill, 'Individual Differences in Developmental Precision and Fluctuating Asymmetry: A Model and its Implications', *Journal of Evolutionary Biology*, 12 (1999), pp. 402–16; R. Thornhill and A. P. Møller, 'Developmental Stability, Disease and Medicine', *Biological Reviews*, 72 (1997), pp. 497–548.
9. A. P. Møller, 'Developmental Stability and Fitness: A Review', *American Naturalist*, 149 (1997), pp. 916–32.
10. A. R. Palmer and C. Strobeck, 'Fluctuating Asymmetry: Measurement, Analysis, Patterns', *Annual Review of Ecology and Systematics*, 17 (1986), pp. 391–421; A. R. Palmer and C. Strobeck, 'Fluctuating Asymmetry as a Measure of Developmental Stability: Implications of Non-Normal Distributions and Power of Statistical Tests', *Acta Zoologica Fennica*, 191 (1992), pp. 57–72; R. Thornhill and A. P. Møller, 'Developmental Stability, Disease and Medicine', *Biological Reviews*, 72 (1997), pp. 497–548.
11. P. A. Parsons, 'Fluctuating Asymmetry: A Biological Monitor of Environmental and Genomic Stress', *Heredity*, 68 (1992), pp. 361–4.
12. D. A. Almeida, G. G. Almodóvar, Nicola and B. Elvira, 'Fluctuating Asymmetry, Abnormalities and Parasitism as Indicators of Environmental Stress in Cultured Stocks of Goldfish and Carp', *Aquaculture*, 279 (2008), pp. 120–5.
13. J. A. Hódar, 'Leaf Fluctuating Asymmetry of Holm Oak in Response to Drought Under Contrasting Climatic Conditions', *Journal of Arid Environments, 52* (2002), pp. 233–43.
14. A. P. Møller, G. S. Sanotra and K. S. Vestergaard, 'Developmental Instability and Light Regime in Chickens (*Gallus gallus*)', *Applied Animal Behaviour Science*, 62 (1999), pp. 57–71; J. W. Chapman and D. Goulson, 'Environmental versus genetic influences on fluctuating asymmetry in the house fly, *Musca domestica*', *Biological Journal of the Linnean Society*, 70 (2000), pp. 403–13.
15. R. Garland and P. W. Freeman, 'Selective Breeding for High Endurance Running Increases Hindlimb Symmetry', Evolution, 59 (2005), pp. 1851–54; P. Galeotti, R. Sacchi and V. Vicario, 'Fluctuating Asymmetry in Body Traits Increases Predation Risks: Tawny Owl Selection Against Asymmetric Woodmice', Evolutionary Ecology, 19 (2005), pp. 405–18; W. U. Blanckenhorn, U. Kraushaar and C. Reim, 'Sexual Selection

on Morphological and Physiological Traits and Fluctuating Asymmetry in the Yellow Dung Fly', *Journal of Evolutionary Biology*, 16 (2003), pp. 903–13.

16. P. T. Rintamäki, R. V. Alatalo, J. Höglund and A. Lundberg, 'Fluctuating Symmetry and Copulation Success in Lekking Black Grouse', *Animal Behaviour*, 54 (1997), pp. 265–9.
17. W. U. Blanckenhorn, U. Kraushaar and C. Reim, 'Sexual Selection on Morphological and Physiological Traits and Fluctuating Asymmetry in the Yellow Dung Fly', *Journal of Evolutionary Biology*, 16 (2003), pp. 903–13.
18. J. P. Swaddle, 'Within-Individual Changes in Developmental Stability Affect Flight Performance', *Behavioral Ecology*, 8 (1997), pp. 601–4.
19. G. R. Allen and L. Simmons, 'Coercive Mating, Fluctuating Asymmetry and Male Mating Success in the Dung Fly, *Sepsis cynipsea*', *Animal Behaviour*, 52 (1996), pp. 737–41.
20. L. W. Simmons, 'Correlates of Male Quality in Field Cricket, *Gryllus Campestris L.*: Age, Size and Symmetry Determine Pairing Success in Field Populations', *Behavioral Ecology*, 6 (1995), pp. 376–81.
21. J. W. Chapman and D. Goulson, 'Environmental versus genetic influences on fluctuating asymmetry in the house fly, *Musca domestica*', *Biological Journal of the Linnean Society*, 70 (2000), pp. 403–13.
22. K. Grammar and R. Thornhill, 'Human (*Homo sapiens*) Facial Attractiveness and Sexual Selection: The Role of Symmetry and Averageness', *Journal of Comparative Psychology*, 108 (1994), pp. 233–42.
23. B. C. Jones, A. C. Little, I. S. Penton-Voak, B. P. Tiddeman, D. M. Burt and D. I. Perrett, 'Facial Symmetry and Judgements of Apparent Health: Support for "Good Genes" Explanation of Attractiveness-Symmetry Relationship', *Evolution and Human Behavior*, 22 (2001), pp. 417–29.
24. R. Thornhill and S. W. Gangestad, 'Facial Attractiveness', *Trends in Cognitive Sciences, 3* (1999), pp. 452–60; B.C. Jones, A. C. Little, I. S. Penton-Voak, B. P. Tiddeman, D. M. Burt and D. I. Perrett, 'Facial Symmetry and Judgements of Apparent Health: Support for "Good Genes" Explanation of Attractiveness-Symmetry Relationship', *Evolution and Human Behavior*, 22 (2001), pp. 417–29; K. Grammar and R. Thornhill, 'Human (*Homo sapiens*) Facial Attractiveness and Sexual Selection: The Role of Symmetry and Averageness', *Journal of Comparative Psychology*, 108 (1994), pp. 233–42.
25. S. Van Dongen, R. Cornille and L. Lens, 'Sex and Asymmetry in Humans: What is the Role of Developmental Instability?', *Journal of Evolutionary Biology*, 22 (2009), pp. 612–22.
26. S. M. Hughes, M. A. Harrison and G. G. Gallup, 'The Sound of Symmetry: Voice as a Marker of Developmental Instability', *Evolution and Human Behaviour*, 23 (2002), pp. 173–80.
27. R. Thornhill and S. W. Gangestad, 'Human Fluctuating Asymmetry and Sexual Behavior', *Psychological Science*, 5 (1994), pp. 297–302.
28. W. M. Brown, M. E. Price, J. Kang, N. Pound, Y. Zhao and H. Yu, 'Fluctuating Asymmetry and Preferences for Sex-Typical Bodily Characteristics', *Proceedings of the National Academy of Sciences of the United States of America*, 105 (2008), pp. 12938–43.
29. S. M. Martin, J. T. Manning and C. F. Dowrick, 'Fluctuating Asymmetry, Relative Digit Length, and Depression in Men', *Evolution and Human Behavior*, 20 (1999), pp. 203–14.
30. J. T. Manning, 'Fluctuating Asymmetry and Body Weight in Men and Women: Implications for Sexual Selection', *Ethology and Sociobiology*, 16 (1995), pp. 145–3.

31. R. M Malina and P. H. Buschang, 'Anthropometric Symmetry in Normal and Mentally Retarded Males', *Annals of Human Biology, 11* (1984), pp. 515–31.
32. R. J. Thoma, R .A. Yeo, S. Gangestad, E. Halgren, J. Davis, K. M. Paulson and J. D. Lewine, 'Developmental Instability and the Neural Dynamics of the Speed-Intelligence Relationship', *NeuroImage,* 32 (2006), pp. 1456–64.
33. T. A. Markow, 'Genetics and Developmental Stability: An Integrative Conjecture on the Etiology and Neurobiology of Schizophrenia', *Psychological Medicine, 22* (1992), pp. 295–305.
34. S. M. Martin, J. T. Manning and C. F. Dowrick, 'Fluctuating Asymmetry, Relative Digit Length, and Depression in Men', *Evolution and Human Behavior,* 20 (1999), pp. 203–14.
35. W. M. Brown, L. Cronk, K. Grochow, A. Jacobson, C. K. Liu, Z. Popović and R. Trivers, 'Dance Reveals Symmetry Especially in Young Men', *Nature,* 438 (2005), pp. 1148–50.
36. J. P. Swaddle, 'Within-Individual Changes in Developmental Stability Affect Flight Performance', *Behavioral Ecology, 8* (1997), pp. 601–4.
37. J. Martin and P. López, 'Hindlimb Asymmetry Reduces Escape Performance in the Lizard *Psammodromus algirus*', *Physiological and Biochemical Zoology, 74* (2001), pp. 619–24.
38. J. T. Manning and L. J. Pickup, 'Symmetry and Performance in Middle Distance Runners', *International Journal of Sports Medicine,* 19 (1998), pp. 205–9.
39. D. Longman, J. T. Stock and J. C. Wells, 'Fluctuating Asymmetry as a Predictor for Rowing Ergometer Performance', *International Journal of Sports Medicine,* 32 (2011), pp. 606–10.
40. E. Al-Eisa, D. Egan and R. Wassersug, 'Fluctuating Asymmetry and Low Back Pain', *Evolution and Human Behaviour,* 25 (2004), pp. 31–7.
41. W. M. Brown, L. Cronk, K. Grochow, A. Jacobson, C. K. Liu, Z. Popović and R. Trivers, 'Dance Reveals Symmetry Especially in Young Men', *Nature,* 438 (2005), pp. 1148–50.
42. W. M. Brown and C. Moore, 'Fluctuating Asymmetry and Romantic Jealousy', *Evolution and Human Behavior,* 24 (2003), pp. 113–17.
43. W. M. Brown and C. Moore, 'Fluctuating Asymmetry and Romantic Jealousy', *Evolution and Human Behavior,* 24 (2003), pp. 113–17.
44. J. R. Kellner and R. A. Alford, 'The Ontogeny of Fluctuating Asymmetry', *The American Naturalist,* 161 (2003), pp. 931–47.
45. A. P. Møller and R. Thornhill, 'Bilateral Symmetry and Sexual Selection: A Meta-Analysis' , *American Naturalist,* 151(1998), pp. 174–92.
46. T. A. Sangster, A. Bahrami, A. Wilczek, E. Watanabe, K. Schellenberg, C., McLellan, A. Kelley, A. Won Kong, C. Queitsch and S. Lindquist, 'Phenotypic Diversity and Altered Environmental Plasticiy in *Arabidopsis thaliana* with Reduced *Hsp*90 Levels', *PloS ONE,* 2 (2007), pp. 1–15; K. H. Takahashi, P. J. Daborn, A. A. Hoffmann and T. Takano-Shimizu, 'Environmental Stress-Dependent Effects of Deletions Encompassing *Hsp70Ba* on Canalization and Quantitative Trait Asymmetry in *Drosophilia melanogaster*', *PloS ONE,* 6 (2011), pp. 1–15; C. Vangestel, J. Mergeay, D. A. Dawson, V. Vandomme and L. Lens, 'Developmental Stability Covaries With Genome-Wide and Single-Locus Heterozygosity in House Sparrows', *PloS ONE,* 6(2011), pp. 1–10.
47. C. C. Milton, B. Huynh, P. Batterham, S. L. Rutherford and A. A. Hoffman, 'Quantitative Trait Symmetry Independent of *Hsp*90 Buffering: Distinct Modes of Genetic Canalization and Developmental Stability', *Proceedings of the National Academy of Sciences of the United States of America,* 100 (2003), pp. 13396–401; K. H. Takahashi, L. Rako, T. Takano-Shimizu, A. A. Hoffmann and S. F. Lee, 'Effects of Small *Hsp* Genes on

Developmental Stability and Microenvironmental Canalization', *BioMed Central Evolutionary Biology*, 10 (2010), doi:10.1186/1471-2148-10-284.

48. S. W. Gangestad, L. A. Merriman and M. E. Thompson, 'Men's Oxidative Stress, Fluctuating Asymmetry and Physical Attractiveness', *Animal Behaviour*, 80 (2010), pp. 1005–13.
49. T. A. Markow, 'Evolutionary Ecology and Developmental Instability', *Annual Review of Entomology*, 40 (1995), pp. 105–20.
50. J. E. Strassmann, D. C. Queller, J. C. Avise and F. J. Ayala, 'In the Light of Evolution V: Cooperation and Conflict', *Proceedings of the National Academy of Sciences of the United States of Americ*, 108 (2011), pp. 10787–91.
51. J. H. Werren, 'Selfish Genetic Elements, Genetic Conflict, and Evolutionary Innovation', *Proceedings of the National Academy of Sciences of the United States of America*, 108 (2011), pp. 10863–70.
52. G. D. D. Hurst and J. H. Werren, 'The Role of Selfish Genetic Elements on Eukaryotic Evolution', *Nature Reviews Genetics*, 2 (2001), pp. 597–606.
53. D. A. Hickey, 'Selfish DNA: A Sexually Transmitted Nuclear Parasite', Genetics 101 (1982), pp. 519–31.
54. M. R. Goddard, D. Greig and A. Burt, 'Outcrossed Sex Allows a Selfish Gene to Invade Yeast Populations', *Proceedings of Royal Society B Biological Sciences*, 268 (2001), pp. 2537–42.
55. M. R. Goddard, D. Greig and A. Burt, 'Outcrossed Sex Allows a Selfish Gene to Invade Yeast Populations', *Proceedings of Royal Society B Biological Sciences*, 268 (2001), pp. 2537–42.
56. L. A. Vøllestad, K. Hindar and A. P. Møller, 'A Meta-Analysis of Fluctuating Asymmetry in Relation to Heterozygosity', *Heredity*, 83 (1999), pp. 206–18.
57. Ibid.
58. Ibid.
59. D. C. Bittel, N. Kibiryeva and S. G. McNulty et al., 'Whole Genome Microarray Analysis of Gene Expression in an Imprinting Center Deletion Mouse Model of Prader-Willi Syndrome', *American Journal of Medical Genetics*, 143A (2007), pp. 422–9; B. Cannon and J. Nedergaard, 'Brown Adipose Tissue: Function and Physiological Significance', *Physiological Reviews*, 84 (2004), pp. 277–359; S. T. da Rocha, M. Tevendale, E. Knowles et al., Restricted Co-Expression of *Dlk1* and the Reciprocally Imprinted Non-Coding RNA, *Gtl2*: Implications for Cis-Acting Control', *Developmental Biology*, 306 (2007), pp. 810–23; D. Haig, 'Huddling: Brown Fat, Genomic Imprinting and the Warm Inner Glow', *Current Biology*, 18 (2008), pp. R172–R175.
60. D. Haig, 'Huddling: Brown Fat, Genomic Imprinting and the Warm Inner Glow', *Current Biology*, 18 (2008), pp. R172–R175.
61. See A. E. Kammer and B. Heinrich, 'Metabolic Rates Related to Muscle Activity in Bumblebees', *Journal of Experimental Biology*, 6 (1974), pp. 219–27.
62. Vøllestad et al., 'A Meta-Analysis of Fluctuating Asymmetry in Relation to Heterozygosity'.
63. L. Keller and L. Passera, 'Incest Avoidance, Fluctuating Asymmetry, and the Consequences of Inbreeding in *Iridomyrmex humilis*, an Ant with Multiple Queen Colonies', *Behavioral Ecology and Sociobiology*, 33 (1993), pp. 191–9.
64. L. J. Leamy, S. Meagher, S. Taylor, L. Carroll and W. K. Potts, 'Size and Fluctuating Asymmetry of Morphometric Characters in Mice: Their Associations with Inbreeding and T-Haplotype', *Evolution*, 55 (2001), pp. 2333–41.

65. J. Blagojević and M. Vujosević, 'B Chromosomes and Developmental Homeostasis in the Yellow-Necked Mouse, *Apodemus flavicollis* (*Rodentia, Mammalia*): Effects on Non-metric Traits', *Heredity*, 93 (2004), pp. 249–54.
66. J. Blagojevićand M. Vujosević, 'B Chromosomes and Developmental Homeostasis in the Yellow-Necked Mouse, *Apodemus flavicollis* (*Rodentia, Mammalia*): Effects on Nonmetric Traits', *Heredity*, 93 (2004), pp. 249–54.
67. J. A. Zeh and D. W. Zeh, 'The Evolution of Polyandry I: Intragenomic Conflict and Genetic Incompatibility', *Proceedings of the Royal Society of London Series B – Biological Sciences*, 263 (1996), pp. 1711–17.
68. Ibid.
69. D. Hasselquist and P. W. Sherman, Social Mating Systems and Extra-Pair Fertilizations in Passerine Birds', *Behavioral Ecology*, 12 (2001), pp. 457–66.
70. Zeh and Zeh, 'The Evolution of Polyandry I'.
71. J. J. Cuervo and A. P. Møller, 'Phenotypic Variation and Fluctuating Asymmetry in Sexually Dimorphic Feather Ornaments in Relation to Sex and Mating System', *Biological Journal of the Linnean Society*, 68 (1999), pp. 505–29.
72. Zeh and Zeh, 'The Evolution of Polyandry I'.
73. J. J. Cuervo and A. P. Møller, 'Phenotypic Variation and Fluctuating Asymmetry in Sexually Dimorphic Feather Ornaments in Relation to Sex and Mating System', *Biological Journal of the Linnean Society*, 68 (1999), pp. 505–29.
74. Zeh and Zeh, 'The Evolution of Polyandry I'.
75. M. Kawahara and T. Kono, 'Longevity in Mice Without a Father', *Human Reproduction*, 25 (2010), pp. 457–61.
76. Ibid.
77. A. P. Møller, C. L. Fincher and R. Thornhill, 'Why Men Have Shorter Lives Than Women: Effects of Resource Availability, Infectious Disease, and Senescence', *American Journal of Human Biology*, 21 (2009), 357–64.
78. W. M. Brown, 'Parental Antagonism Theory of Language Evolution: Preliminary Evidence for the Proposal', *Human Biology*, 83 (2011), pp. 213–45.
79. D. Haig, 'On Intrapersonal Reciprocity', *Evolution and Human Behavior*, 24 (2003), pp. 418–25.

11 Johnston, 'Beauty, Bacteria and the Faustian Bargain'

1. S. J. Gould, 'Planet of the Bacteria', *Washington Post Horizon*, 119 (1996), p. 344.
2. W. B. Whitman, D. C. Coleman and W. J. Wiebe, 'Prokaryotes: The Unseen Majority', *Proceedings of the National Academy of Sciences USA*, 95 (1998), pp. 6578–83.
3. W. Osler, 'The Study of the Fevers of the South', *Journal of the American Medical Association*, 26 (1896), pp. 999–1004.
4. W. D. Hamilton, 'Sex Versus Non-Sex Versus Parasite', *Oikos*, 35 (1980), pp. 282–90. W. D. Hamilton, P. Henderson and N. Moran, 'Fluctuation of Environment and Coevolved Antagonist Polymorphism as Factors in the Maintenance of Sex', in R. D. Alexander and D. W. Tinkle (eds), *Natural Selection and Social Behavior* (New York: Chiron Press, 1981), pp. 363–81.
5. C. G. Williams, *Sex and Evolution* (Princeton NJ: Princeton University Press, 1975).
6. M. Daly and Wilson, M., *Sex, Evolution and Behavior* (Boston, MA: Willard Grant Press, 1983).

7. Hamilton et al. 'Fluctuation of Environment and Coevolved Antagonist Polymorphism as Factors in the Maintenance of Sex'.
8. L. Van Valen, A New Evolutionary Law', *Evolutionary Theory*, 1 (1973), pp. 1–30.
9. E. Walster, V. Aronson, D. Abrahams and L. Rottmann, 'Importance of Physical Attractiveness in Dating Behavior', *Journal of Personality and Social Psychology*, 4 (1966), pp. 508–16.
10. J. H. Langlois and L. A. Roggman, 'Attractive Faces are Only Average', *Psychological Science*, 1 (1990), pp. 115–21; J. H. Langlois, L. A. Roggman and L. Musselman, 'What is Average and What is not Average about Attractive Faces?', *Psychological Science*, 5 (1994), pp. 214–20.
11. T. R. Alley and M. R. Cunningham, 'Averaged Faces are Attractive, but Very Attractive Faces are not Average', *Psychological Science*, 2 (1991), pp. 123–5; R. Thornhill and S. W. Gangestad, 'Human Facial Beauty: Averageness, Symmetry, and Parasite Resistance', *Human Nature*, 4 (1993), pp. 237–69; V. S. Johnston, 'Female Facial Beauty: The Fertility Hypothesis', *Pramatics and Cognition*, 8 (2000), pp. 107–22.
12. V. S. Johnston, D. R. Miller and M. Franklin, 'Is Beauty in the Eye of the Beholder?' *Ethology and Sociobiology*, 14 (1993), pp. 183–99; D. I. Perrett, K. A. May and S. Yoshikawa, 'Facial Shapes and Judgments of Female Attractiveness', *Nature*, 368 (1994), pp. 239–42; D. I. Perrett, K. A. May, S. Yoshikawa, K. J. Lee, I. Penton-Voak, D. Rowland, S. Yoshikawa, D. M. Burt, S. P.Henzi, D. L. Castles and S. Akamatsu, 'Effects of Sexual Dimorphism on Facial Attractiveness', *Nature*, 394 (1998), pp. 884–7.
13. M. R. Cunningham, A. R. Roberts, A. P. Barbee, P. B. Druen, and Cheng-Huan Wu, 'Their Ideas of Beauty are, on the Whole, the Same as Ours: Consistency and Variability in the Cross-cultural Perceptions of Female Physical Attractiveness', *Journal of Personality and Social Psychology*, 68 (1995), pp. 261–79.
14. G. B. Forbes, 'Puberty: Body Composition', in S. R. Berenson (ed.), *Puberty* (Leiden: Stenfert-Kroese, 1975), pp. 132–45; J. M. Tanner, '*Fetus into Man: Physical Growth from Conception to Maturity*' (Cambridge, MA: Harvard University Press, 1990); M. M. Grumbach and R. J. Auchus 'Estrogen: Consequences and Implications of Human Mutations in Synthesis and Action', *Journal of Clinical Endocrinology and Metabolism*, 84 (1999), pp. 4677–94.
15. D. Singh, 'Body Shape and Woman's Attractiveness: The Critical Role of Waist-to-hip Ratio', *Human Nature*, 4 (1993), pp. 297–321; B. M. Zaadstra, J. C Seidell, P. A. Van Noord, E. R. te Velde, J. D. Habbema, B. Vrieswijk, and J. Karbaat, 'Fat and Female Fecundity: Prospective Study of Effect of Body Fat Distribution and Conception Rates', *British Medical Journal*, 306 (1993), pp. 484–7.
16. L. Henry, 'Some Data on Natural Fertility', *Eugenics Quarterly*, 8 (1961), pp. 81–91.
17. V. S. Johnston, D. R. Miller and M. H. Burleson, 'Multiple P3s to Emotional Stimuli and their Theoretical Significance', *Psychophisiology*, 23 (1986), pp. 684–94.
18. V. S. Johnston, D. R. Miller, M. H. Burleson and J. C. Oliver-Rodriguez, 'Facial Beauty and the Late Positive Component of Event-related Potentials', *Journal of Sex Research*, 34 (1996), pp. 188–98.
19. S.B. Hrdy and P. L. Whitten, 'Patterning of Sexual Activity', in B. B. Smuts, D. L. Cheney, R. M. Seyfarth, R. W. Wrangham and T. T. Struhsaker (eds.), *Primate Societies* (Chicago, IL: University of Chicago Press, 1987), pp. 370–84.
20. M. B. Andersson, 'Female Choice Selects for Extreme Tail Length in a Widowbird', *Nature*, 299 (1982), pp. 818–20.

21. M. Petrie, 'Improved Growth and Survival of Offspring of Peacocks with more Elaborate Trains', *Nature*, 371 (1994), pp. 598–9.
22. V. S. Johnston, D. R. Miller, M. H. Burleson, R. Hagel, M. Franklin, B. Fink and K. Grammer, 'Male Facial Attractiveness: Evidence for Hormone Mediated Adaptive Design', *Evolution and Human Behavior*, 22 (2001), pp. 251–67.
23. I. S. Penton-Voak, D. I. Perrett, D. L. Castles, T. Kobayashi, D. M. Burt, L. K. Murray and R. Minamisawa, 'Menstrual Cycle Alters Face Preference', *Nature*, 399 (1999), pp. 741–2; P. Scarbrough and V. S. Johnston, 'Individual Differences in Women's Facial Preferences as a Function of Digit Ratio and Mental Rotation Ability', *Evolution and Human Behavior*, 26 (2005), pp. 509–26.
24. L. Van Valen, 'A Study of Fluctuating Asymmetry', *Evolution*, 16 (1962), pp. 125–42.
25. A. P. Møller and R. Thornhill, 'Bilateral Symmetry and Sexual Selection: A Meta-Analysis', *American Naturalist*, 151 (1997), pp. 174–92; R. Thornhill and S. W. Gangestad, 'Human Fluctuating Asymmetry and Sexual Behavior', *Psychological Science*, 5 (1994), pp. 297–302; D. Waynforth, 'Fluctuating Asymmetry and Human Male Life-History Trait in Rural Belize', *Proceeding of the Royal Society of London*, 265 (1998), pp. 1497–501.
26. S. Gangestad, R. Thornhill and R. A. Yeo, 'Facial Attractiveness, Developmental Stability and Fluctuating Asymmetry', *Ethology and Sociobiology*, 15 (1994), pp. 73–85
27. S. Gangestad, R. Thornhill and R. A. Yeo, 'Facial Masculinity and Fluctuating Asymmetry', *Evolution and Human Behavior*, 24 (2003), pp. 231–41.
28. V. S. Johnston, *Why we Feel: The Science of Human Emotions* (Reading, MA: Perseus Press, 1999).
29. Ibid.
30. N. Lane, *Life Ascending: The Ten Great Inventions of Evolution* (London: W. W. Norton & Company Ltd, 2010), p. 124.

12 Engels, 'Darwin's Care for Humanity'

1. See also E.-M. Engels, 'Darwin's Philosophical Revolution: Evolutionary Naturalism and First Reactions to his Theory', in E.-M. Engels and Th. F. Glick (eds), *The Reception of Charles Darwin in Europe*, 2 vols (London and New York: Continuum, 2008), pp. 23–53. For a more extensive presentation of Darwin's evolutionary anthropology and of his ethics see my German monograph E.-M. Engels, *Charles Darwin* (München: C. H. Beck, 2007).
2. C. Darwin, *The Origin of Species* (1859; Reprint Cambridge, MA and London: Harvard University Press, 1964).
3. N. Barlow (ed.), *The Autobiography of Charles Darwin 1809–1882* (1958; New York and London: W. W. Norton & Company, 1969), p. 122.
4. Darwin, *Origin of Species*, p. 459. From the sixth edition of 1872 on, his last edition, he names it 'theory of descent with modification through variation and natural selection', although variation was from the very beginning on a central element of his theory.
5. Ibid., p. 30.
6. Ibid., p. 471.
7. T. R. Malthus, *An Essay on the Principle of Population*, 2 vols, The version published in 1803, with the variora of 1806, 1807, 1817 and 1826, ed. P. James. (Cambridge: Cambridge University Press, 1989), vol. 1, p. 15. Malthus drew on Benjamin Franklin who applied this principle also to plants and animals.
8. Ibid., vol. 1, pp. 16–18.

9. Ibid., vol. 1, p. 18, footnote 4.
10. Ibid., vol. 1, p. 23.
11. Ibid., vol. 2, p. 96.
12. Darwin, *The Origin of Species*, p. 63.
13. H. Spencer, *The Principles of Biology*, 2 vols (1864; Osnabrück: Zeller, 1966), vol. 1, pp. 444–5. The term 'survival of the fittest' does not mean 'survival of the survivor' – this would be a tautology – but it means, that there is an equilibrium between an organism and its environment.
14. Darwin, *The Origin of Species*, p. 64.
15. Ibid., p. 62.
16. C. Darwin, 'To William Preyer, March 29, 1869', in *The Correspondence of Charles Darwin*, ed. F. Burkhardt†, J. A. Secord, S. A. Dean, S. Evans, S. Innes, A. M. Pearn, P. White, 19 of 30 planned vols published (Cambridge: Cambridge University Press, 1985–), vol. 17 (2009), pp. 161–2, on p. 161.
17. D. Todes, *Darwin without Malthus. The Struggle for Existence in Russian Evolutionary Thought* (New York and Oxford: Oxford University Press, 1989).
18. Darwin, *The Origin of Species*, p. 488.
19. P. H. Barrett, P. J. Gautrey, S. Herbert, D. Kohn, S. Smith (eds), *Charles Darwin's Notebooks, 1836–1844* (Ithaca, NY: Cornell University Press, 1987).
20. I am using the second edition of 1874, 1877. C. Darwin (1877) *The Descent of Man, and Selection in Relation to Sex* (1877), 2nd edn (1874; rev. augm. 1877), ed. P. H. Barrett, R. B. Freeman, *The Works of Charles Darwin* (1871), 29 vols (London: William Pickering, 1986–9), vols 21, 22 (1989).
21. Darwin, *The Descent of Man, and Selection in relation to Sex*, vol. 21, p. 101.
22. Ibid., vol. 21, p. 130, see also pp. 69–70.
23. Ibid., vols 21 and 22.
24. Ibid., vol. 21, p. 104.
25. Ibid., vol. 21, p. 103.
26. Ibid., vol. 21, p. 109.
27. Ibid., vol. 21, pp. 109–10.
28. Ibid., vol. 21, p. 111.
29. Ibid., vol. 21, p. 110.
30. Ibid., vol. 21, p. 123.
31. Ibid.
32. Ibid., vol. 21, p. 116.
33. Ibid., vol. 22, p. 637.
34. Ibid., vol. 21, p. 117, footnote 27.
35. Ibid., vol. 21, p. 52.
36. Darwin and Wallace had conceived the same general theory of natural selection independently of each other, Darwin already since 1838 including the human being as object of his explanation, the fourteen year younger Wallace since about 1855. In order to avoid conflicts of priority their papers were read in front of the Linnean Society in 1858. But Wallace *published* his ideas on the application of the general theory to the human being before Darwin: A. R. Wallace, 'The Origin of Human Races and the Antiquity of Man deduced from the theory of "Natural Selection"', *Journal of the Anthropological Society of London*, 2 (1864), pp. clviii–clxxxvii.
37. Ibid., p. clxvi.
38. Ibid.

39. Ibid., p. clxviii.
40. Ibid.
41. Darwin, *The Descent of Man, and Selection in Relation to Sex*, vol. 21, p. 135.
42. Ibid., vol. 21, p. 133, footnote 2.
43. Ibid., vol. 21, p. 135.
44. Ibid., vol. 21, p. 136.
45. Ibid.
46. Ibid., vol. 21, p. 137.
47. Ibid., vol. 21, p. 114.
48. Ibid., vol. 22, p. 637.
49. Ibid., vol. 21, p. 125.
50. Ibid., vol. 21, p. 127.
51. Ibid.
52. Ibid., vol. 21, p. 129.
53. Ibid., vol. 21, p. 130.
54. W. E. H. Lecky, *History of European Morals from Augustus to Charlemagne* (London: Longmans, Green, & Co. 1869), p. 103. Darwin refers however to another page.
55. Darwin, *The Descent of Man, and Selection in relation to Sex*, vol. 21, p. 127.
56. Ibid., vol. 21, p. 127.
57. Ibid., vol. 21, p. 130.
58. Ibid., vol. 22, p. 637.
59. Ibid., vol. 22, p. 643.
60. Ibid., vol. 21, pp. 141–2.
61. Wallace, 'The Origin of Human Races and the Antiquity of Man'; W. R. Greg, 'On the Failure of "Natural Selection" in the Case of Man', *Fraser's Magazine*, September (1868), pp. 353–62; F. Galton, 'Hereditary Talent and Character', *MacMillan's Magazine*, 12 (1865), Part I, pp. 157–66; Part II, pp. 318–27; *Hereditary Genius. An Inquiry into its Laws and Consequences* (1869), 2nd edn (London: Macmillan and Co., 1870).
62. Greg, 'On the Failure of "Natural Selection" in the Case of Man', p. 358.
63. Ibid., p. 356.
64. Ibid., p. 358.
65. Ibid.
66. Ibid., p. 360.
67. Ibid., pp. 360–1. Compare Plato, *Republic*, trans. R. Waterfield (Oxford and New York: Oxford University Press, 1993), book V, 457b-461e.
68. Ibid., p. 362.
69. Ibid., p. 359.
70. Ibid.
71. Ibid., p. 362.
72. Galton, 'Hereditary Talent and Character', p. 165.
73. F. Galton, *Memories of My Life* (London: Methuen & Co, 1908), p. 287.
74. Galton, 'Hereditary Talent and Character', p. 165.
75. Ibid., pp. 319–20.
76. C. P. Blacker, *Eugenics in Retrospect and Prospect. The Galton Lecture, 1945* (1945), 2nd edn (London: Eugenics Society and Cassell & Company, 1950), p. 21.
77. F. Galton, *Inquiries into Human Faculty and its Development* (London: Macmillan & Co, 1883), p. 25.
78. Galton, 'Hereditary Talent and Character', p. 326.
79. Ibid.

80. F. Galton, 'Eugenics: Its Definition, Scope and Aims', *Sociological Papers*, 1 (1905), pp. 45–51, on p. 50.
81. Ibid., p. 50.
82. Galton, *Memories of My Life*, p. 323.
83. Ibid.
84. Ibid.
85. Darwin, *The Descent of Man, and Selection in Relation to Sex*, vol. 21, p. 139.
86. Ibid.
87. Ibid.
88. I thank Prof. Dr Richard Burian, Dr Joy Harvey and Walter H. Nilson for their assistance with translating and interpreting this section.
89. S. F. Weiss, 'The Race Hygiene Movement in Germany, 1904–1945', in M. B. Adams (ed.), *The Wellborn Science. Eugenics in Germany, France, Brazil, and Russia* (New York and Oxford: Oxford University Press, 1990), p. 16.
90. Darwin, *The Descent of Man, and Selection in Relation to Sex*, vol. 21, p. 139.
91. Ibid., vol. 22, p. 643.
92. Ibid.
93. G. H. Darwin, 'MARRIAGES *between* FIRST COUSINS *in* ENGLAND *and their* EFFECTS', *Journal of the Statistical Society*, 38 (June 1875), pp. 153–84.
94. Ibid., p. 183.
95. F. Darwin, 'Francis Galton. 1822–1911', *Eugenics Review*, 6:1 (1914), pp. 1–17, on p. 16.
96. C. Darwin, 'To Francis Galton, Down, Jan. 4th, 1873', in *More Letters of Charles Darwin*, ed. F. Darwin, 2 vols (London: John Murray, 1903), vol. 2, pp. 43–4.
97. C. Darwin, *Variation of Animals and Plants under Domestication*, 2nd rev. edn (1875), 2 vols, ed. P. H. Barrett, R. B. Freeman, in *The Works of Charles Darwin* (1868), 29 vols (London: William Pickering, 1986–9), vols 19, 20 (1988), vol. 20, p. 153.
98. C. P. Blacker, *Eugenics in Retrospect and Prospect. The Galton Lecture, 1945*, p. 8.
99. Darwin, *The Descent of Man, and Selection in Relation to Sex*, vol. 21, p. 141.
100. Ibid.
101. Ibid., vol. 21, p. 143.
102. Ibid., vol. 21, pp. 141–2.
103. P. Tort, 'Effet réversif de l'évolution, Reversive effect of evolution', *Dictionnaire du Darwinisme et de l'Évolution*, publié sous la Direction de Patrick Tort, 3 vols (Paris: Presses Universitaires de France, 1996), vol. 1, pp. 1334–5.
104. Darwin, *The Descent of Man, and Selection in Relation to Sex*, vol. 22, p. 643.
105. Contraceptives in the nineteenth century were diaphragms, condoms, sponges, acidic powders and jellies etc. See A. McLaren, *A History of Contraception. From Antiquity to the Present Day* (Oxford: Basil Blackwell, 1990).
106. S. Chandrasekhar (ed.), 'The Life and Work of Knowlton and his FRUITS OF PHILOSOPHY', in S. Chandrasekhar (ed.), *'A Dirty, Filthy Book'. The Writings of Charles Knowlton and Annie Besant on Reproductive Physiology and Birth Control and an Account of the Bradlaugh-Besant Trial* (Berkeley, Los Angeles, CA and London: University of California Press, 1981), pp. 21–5, on p. 22.
107. A. Besant, 'The Law of Population: Its Consequences, and its Bearing upon Human Conduct and Morals' (first published in London: Freethought Publishing Company, 1884), in S. Chandrasekhar (ed.), *'A Dirty, Filthy Book'. The Writings of Charles Knowlton and Annie Besant on Reproductive Physiology and Birth Control and an Account of the Bradlaugh-Besant Trial*, pp. 149–201, on p. 175.
108. Malthus, *An Essay on the Principle of Population*, vol. 2, p. 235.

109. Besant, 'The Law of Population: Its Consequences, and its Bearing upon Human Conduct and Morals', p. 181.
110. Ibid., p. 182.
111. Ibid., p. 184.
112. Ibid.
113. Ibid., pp. 179–80.
114. Ibid., p. 196.
115. See the informative description of the trial and its background by S. Chandrasesekhar, 'The Bradlaugh–Besant Trial, 1877–1878' in his collection of texts: Chandrasekhar (ed.), *'A Dirty, Filthy Book'. The Writings of Charles Knowlton and Annie Besant on Reproductive Physiology and Birth Control and an Account of the Bradlaugh-Besant Trial*, on pp. 26–54.
116. Ibid., p. 46.
117. See the description of the trial in C. R. Drysdale, *The Population Question according to T. R. Malthus and J. S. Mill giving the Malthusian Theory of Over-Population* (London: William Bell, 1880), p. 80; see also A. Desmond and J. Moore, *Darwin* (London: Michael Joseph, 1991), pp. 627–8 and J. Browne, *Charles Darwin. The Power of Place* (New York: Alfred A. Knopf, 2002), pp. 443–5.
118. I thank the Editors of the Darwin Correspondence Project, University of Cambridge, for access to unpublished material, to the correspondence between Ch. Bradlaugh and Ch. Darwin. The quoted material is at a pre-publication stage, and the Project cannot be held responsible for any errors of transcription remaining. I also thank the copyright holder of this letter, Mr. William Huxley Darwin (London), for his permission to quote the letter. Finally I thank Dr. Paul White from the Darwin Correspondence Project for our helpful correspondence on the interpretation of this letter.
119. Darwin, *The Descent of Man, and Selection in Relation to Sex*, vol. 21, p. 139. H. Fick, 'Ueber den Einfluss der Naturwissenschaft auf das Recht', *Jahrbücher für Nationalökonomie und Statistik*, 18 (1872), pp. 248–77.
120. R. Weikart, 'A Recently Discovered Darwin Letter on Social Darwinism', *Isis*, 86 (1995), pp. 609–11, on p. 611.
121. Ibid., p. 611.
122. Besant, 'The Law of Population: Its Consequences, and its Bearing upon Human Conduct and Morals', pp. 196–7.
123. Darwin, *The Descent of Man, and Selection in Relation to Sex*, vol. 21, p. 139.
124. For reading on the poor laws in general see D. Fraser (ed.), *The New Poor Law in the Nineteenth Century* (London, Basingstoke: The Macmillan Press, 1976); M. E. Rose, *The Relief of Poverty 1834–1914* (London, Basingstoke: The Macmillan Press, 1972, reprint 1974).
125. Darwin, *The Descent of Man, and Selection in Relation to Sex*, vol. 22, p. 643.
126. Ibid., vol. 21, p. 147.
127. G. A. Gaskell, 'To Charles Darwin, November 13th, 1878', in J. H. Clapperton, *Scientific Meliorism and the Evolution of Happiness* (London, Beccles: William Clowes and Sons, 1885, reprint Bibliobazaar), pp. 337–340, on p. 337.
128. Ibid., p. 338.
129. Gaskell, 'To Charles Darwin, November 20th, 1878', in Clapperton, *Scientific Meliorism and the Evolution of Happiness*, pp. 341–2, on p. 342.
130. Gaskell, 'To Charles Darwin, November 13th, 1878', p. 339.

131. Ibid.
132. Ibid., p. 340.
133. C. Darwin, 'To G. A. Gaskell, November 15th, 1878', in Clapperton, *Scientific Meliorism and the Evolution of Happiness*, pp. 340–1, on p. 340.
134. Ibid., p. 340.
135. Ibid., pp. 340–1.
136. Darwin, *The Descent of Man, and Selection in Relation to Sex*, vol. 21, p. 121.
137. Ibid., vol. 22, p. 629.
138. Ibid., vol. 21, p. 148.
139. F. Galton, *Memories of My Life*, p. 287.

13 De Sousa, 'Reproduction and Social Selection: The Eugenics Maelstrom of Science, Intelligentsia and Reformers'

1. K. Pearson, *The Life, Letters and Labours of Francis Galton, vol. 2: Researches of Middle Life, Eugenics* (Cambridge: Cambridge University Press, 1924), pp. 170–1.
2. F. Galton, *Inquiries into Human Faculty and its Development* (London: J. M. Dent and Sons, 1883), p. 336.
3. Pearson, *The Life, Letters and Labours of Francis Galton, vol. 2: Researches of Middle Life*, p. 249.
4. Ibid., p. 249.
5. K. Pearson, *The Life, Letters and Labours of Francis Galton, vol. 3. Part A: Correlation, Personsl Identification and Eugenics* (Cambridge: Cambridge University Press, 1930), p. 93.
6. Galton, *Inquiries into Human Faculty and its Development*, p. 241.
7. K. Pearson, *The Grammar of Science* (London: Walter Scott, 1892), p. 33.
8. K. Pearson, *The Groundwork of Eugenics* (London: Dulau 1909), p. 21; S. Szreter, *Fertility, Class and Gender in Britain 1860–1940* (Cambridge: Cambridge University Press, 1996), p. 132.
9. F. C. S. Schiller, 'National Self-Selection', *Eugenics Review*, 2:1 (1910), pp. 8–24, on p. 8.
10. Galton, *Inquiries into Human Faculty and its Development*, p. 333.
11. J. Slaughter, 'Selection in Marriage', *Eugenics Review*, 1:3 (1909), pp. 150–162, on p. 151.
12. F. Galton, 'Eugenic Qualities of Primary Importance', *Eugenics Review*, 1:2 (1909), pp. 74–6, on pp. 74, 76.
13. F. Galton, 'Hereditary Talent and Character', *Macmillan's Magazine*, 12 (1865), pp. 318–27, on p. 320.
14. See E. Barkan, *The Retreat of Scientific Racism* (Cambridge: Cambridge University Press, 1992); and G. Schaffer, *Racial Science and British Society, 1930–62* (Basingstoke: Palgrave MacMillan, 2008).
15. J. Waller, 'Ideas of Heredity, Reproduction and Eugenics in Britain, 1800–1875', *Studies in History Philosophy Biology&Biomedical Sciences*, 32:3 (2001), pp. 457–89.
16. F. Galton 'Foreword', *Eugenics Review*, 1:2 (1909), pp.1–2, on p.1.
17. F. Schenk and A. S. Parkes, 'The Activities of the Eugenics Society', *Eugenics Review*, 60:1 (1968), pp. 142–61.
18. See L. Farrall, *The Origins and Growth of the English Eugenics Movement, 1865–1965* (New York and London: Garland, 1985).
19. P. Mazundar, *Eugenics, Human Genetics and Human Failings* (London: Routledge, 1992), p. 87.

20. A. Carr-Saunders, E. J. Greenwood, E. Lidbetter and A. F. Tredgold, 'The Standardisation of Pedigrees', *Eugenics Review*, 4:4 (1913), pp. 383–90; and E. Lidbetter, 'Nature and Nurture – A Study in Conditions', *Eugenics Review*, 4:1 (1912), pp. 54–73.
21. R. Fisher, 'Some Hopes of a Eugenist', *Eugenics Review*, 5 (1914), pp. 309–15, on p. 311.
22. See W. B. Provine, *The Origins of Theoretical Population Genetics* (Chicago, IL: University of Chicago Press, 1971).
23. B. Bateson, *William Bateson, Naturalist* (Cambridge: Cambridge University Press, 1928), p.305.
24. Ibid., p. 372.
25. Ibid., p. 341.
26. R. Punnett, 'Eliminating Feeblemindedness', *Journal of Heredity*, 8:10 (1917), pp. 464–5, on p. 465.
27. J. B. S. Haldane, *The Inequality of Man and Other Essays* (London: Chatto, 1932).
28. Mazundar, *Eugenics, Human Genetics and Human Failings*, pp. 216–20.
29. L. Farrall, 'The History of Eugenics: A Bibliographical Review', *Annals of Science* 36 (1979), pp. 11–123, on p. 116.
30. H. Laski, 'The Scope of Eugenics', *Westminster Review*, 174 (1910), pp. 25–34, on p. 34.
31. W. McDougall, 'Psychology in the Service of Eugenics', *Eugenics Review*, 5:4 (1914), pp. 295–308, on, p. 306.
32. C. Burt, 'The Inheritance of Mental Characters', *Eugenics Review*, vol. 4:2 (1912), pp. 169–200, on p. 200.
33. C. Spearman, 'The Heredity of Abilities', *Eugenics Review*, 6:3 (1914), pp. 219–37, on p. 234.
34. J. M. Keynes, 'Some Economic Consequences of a Declining Population', *Eugenics Review*, 29:1 (1937), pp. 3–12.
35. J. Toye, 'Keynes on Population and Economic Growth', *Cambridge Journal of Economics*, 21 (1997), pp. 1–26, on p. 6.
36. A. Pigou, *The Economics of Welfare* (London: Macmillan, 1920), p. 112.
37. A. Pigou, 'Social Improvement and Modern Biology', *Economic Journal*, 17:3 (1907), pp. 358–69, on p. 363.
38. Ibid.
39. A. O. Bell, *The Diary of Virginia Woolf, Volume One: 1915–1919* (New York and London: Harcourt Brace Janovich, 1977), p. 13.
40. A. Richardson, *Love and Eugenics in the Late Nineteenth Century* (Oxford: Oxford University Press, 2003), prologue.
41. D. Childs, *Modernism and Eugenics* (Cambridge: Cambridge University Press, 2011), p. 9.
42. H. G. Wells, *Anticipations of the Reaction of Mechanical and Scientific Progress Upon Life and Thought* (1901) (Ithaca: Cornell University Library, 2009), p. 90.
43. Wells, *Anticipations of the Reaction of Mechanical and Scientific Progress Upon Life and Thought*, p. 314.
44. G. R. Searle, 'Eugenics and Class', in C. Webster (ed.), *Biology, Medicine and Society 1840–1940* (Cambridge: Cambridge University Press, 1981), pp. 217–42, on p. 231.
45. Childs, *Modernism and Eugenics*, p. 9.
46. B. Russell, *Marriage and Morals* (1929; New York: Liveright, 1970), p. 259.
47. Ibid.
48. S. Heathorn, 'Explaining Russell's Eugenic Discourse in the 1920s', *Journal of Bertrand Russell Studies*, 25 (Winter 2005–6), pp. 107–39.
49. Childs, *Modernism and Eugenics*, pp. 29–32.

50. See C. S. Jones, *Beatrice Webb. A Life* (Chicago, IL: Ivan R. Dee, 1992).
51. S. Webb, 'Eugenics and the Poor Law', *Eugenics Review*, 2:3 (1910), pp. 233–340, on p. 233.
52. Mazundar, *Eugenics, Human Genetics and Human Failings*, pp. 23–4. Also 'The Feeble-Minded Control Bill: House of Commons Meeting, December 5th 1911', *Eugenics Review*, 3:4 (1912), pp. 355–8.
53. L. Bland and L. Hall, 'Eugenics in Britain: The View from the Metropole', in A. Bashford and P. Levine (eds), *The Oxford Handbook of the History of Eugenics*. (Oxford: Oxford University Press, 2010), pp. 213–27.
54. D. C. Jones, 'Differential Class Fertility', *Eugenics Review*, 24:3 (1932), pp. 175–90, on p. 175.
55. Mazundar, *Eugenics, Human Genetics and Human Failings*, pp. 204–9.
56. Searle, 'Eugenics and Class', p. 223–7.
57. J. Macnicol, 'Eugenics and the Campaign for Voluntary Sterilization in Britain Between the Wars', *Society for the Social History of Medicine*, 2:2 (1989), pp. 147–69, on pp. 151–60.
58. J. Macnicol, 'Eugenics, Medicine and Mental Deficiency: an introduction', *Oxford Review of Education*, 9:3 (1983), pp. 177–80, on p. 178.
59. Macnicol, 'Eugenics and the Campaign for Voluntary Sterilization in Britain between the Wars', pp. 151–62.
60. L. Bland, *Banishing the Beast* (New York: New York Press, 1995), p. 239.
61. H. Ellis, *The Task of Social Hygiene* (London: Constable 1912), p. 186.
62. H. Ellis, 'The Sterilisation of the Unfit', *Eugenics Review*, 1:1 (1909), pp. 203–6, on p. 204.
63. Ibid., p. 206.
64. D. Kevles, *In the Name of Eugenics* (1985; Cambridge, MA: Harvard University Press, 1995), pp. 64–5.
65. See Richardson, *Love and Eugenics in the Late Nineteenth Century*.
66. A. Ravenhill, 'Eugenic Ideals for Womanhood', *Eugenics Review*, 1:4 (1910), pp. 265–4, on p. 273.
67. R. M. Leslie, 'Woman's Progress in relation to Eugenics', *Eugenics Review*, 2:4 (1911), pp. 282–98, on pp. 287, 289.
68. Bland, *Banishing the Beast*, p. 233.
69. L. Darwin, 'Birth Control: A Discussion', *Eugenics Review*, 12:4 (1921), pp. 291–8.
70. R. Soloway, *Demography and Degeneration* (1990; Chapel Hill, CA: University of California Press, 1995), p. 203.
71. M. C. Stopes, *Wise Parenthood* (1918) (London: Putnam's Sons, 1923), p. x.
72. H. G. Wells, 'Discussion of the Galton Eugenics: Its Definition, Scope, and Aims' *American Journal of Sociology*, 10 (1904), pp. 10–11.
73. R. R. Rentoul, *Race Culture; or Race Suicide? A Plea for the Unborn* (1906; New York and London: Garland, 1984), p. 7.
74. E. A. Ross, 'The Causes of Race Superiority', *Annals of the American Academy of Political and Social Science*, 18:1 (1901a), pp. 67–89; *Social Control: A Survey of the Foundations of the Order* (New York: Macmillan, 1901b).
75. T. Roosevelt, 'A Letter from President Roosevelt on Race Suicide', *American Monthly Review of Reviews*, 35:5 (1907), pp. 550–1.
76. Kevles, *In the Name of Eugenics*, p. 104.
77. C. Rosen, *Preaching Eugenics* (Oxford: Oxford University Press, 2004), p. 7.

78. W. Kline, *Building a Better Race* (Berkeley, CA: University of California Press, 2001), pp. 24–7.
79. Kevles, *In the Name of Eugenics*, pp. 90–103.
80. See K. Ludmerer, 'American Geneticists and the Eugenics Movement: 1905–1935', *Journal of the History of Biology*, 2:2 (1969), pp. 337–62.
81. See H. Gruenberg, 'Men and Mice at Edinburgh', *Journal of Heredity*, 30:9 (1939), pp. 371–4.
82. 'The Second International Congress of Eugenics', *Eugenics Review*, 13:1 (1922), pp 512–24.
83. I. Fisher, 'Impending Problems of Eugenics', *Scientific Monthly*, 13:3 (1921), pp. 214–31, on p. 215.
84. Ibid., p. 227.
85. T. Leonard, 'Retrospectives. Eugenics and Economics in the Progressive Era', *Journal of Economic Perspectives*, 19:4 (2005), pp. 207–24, on p. 219.
86. See P. Lombardo (ed.), *A Century of Eugenics in America* (Indianapolis: Indiana University Press, 2011).
87. Kline, *Building a Better Race*, pp. 81, 156.
88. See A. Franks, *Margaret Sanger's Eugenic Legacy. The Control of Female Fertility* (North Carolina: McFarland, 2005).
89. M. Sanger, *The Autobiography of Margaret Sanger* (1938; New York: Dover Publications, 2004), p. 135.
90. Ibid., p. 375.
91. Ibid., p. 377.
92. J. Marks, 'Historiography of Eugenics', *American Journal of Human Genetics*, 52 (1993), pp. 650–2, on p. 651.
93. P. Fara, *A Four Thousand Year History* (Oxford: Oxford University Press, 2009), p. 339.
94. T. Duster, *Backdoor to Eugenics* (London: Routledge, 2003), p. xiv.
95. K. Garver and B. Garver, 'Eugenics: Past, Present, and the Future', *American Journal of Human Genetics*, 49 (1991), pp. 109–18, on p. 109.
96. J. D. Watson, *A Passion for DNA: Genes, Genomes, and Society* (New York: Cold Spring Harbor Laboratory Press, 2000), p. 200.
97. J. Robertson, 'Genetic Selection of Offspring Characteristics', *Boston University Law Review*, 76 (1996), pp. 421–82, on p. 423.
98. R. Aviad, *Community Genetics and Genetic Alliances* (London: Routledge, 2010), p. 12.
99. Duster, *Backdoor to Eugenics*, p. 5.
100. S. J. Peart and D. Levy, *The 'Vanity of the Philosopher'* (Ann Arbor, MI: University of Michigan Press, 2005), p. 4.
101. F. Fukuyama, *Our Posthuman Future* (Great Britain: Profile Books, 2002), p. 87.
102. Ibid., p.107.
103. P. Kitcher, *The Lives to Come* (New York: Simon & Schuster, 1996), p. 201.
104. Ibid.. p. 199.
105. G. Radick, 'A Critique of Kitcher on Eugenic Reasoning', *Studies in History & Philosophy of Biological & Biomedical Sciences*, 32C:4 (2001), pp. 741–51, on p. 744.
106. R. Chadwick, 'What Counts as success in genetic counselling?', *Journal of Medical Ethics*, 19 (1993), pp. 43–6; M. Sandel, *The Case Against Perfection* (Cambridge, MA: Harvard University Press, 2007).
107. J. Harris, 'Is Gene Therapy a Form of Eugenics?', *Bioethics*, 7:2–3 (1993), pp. 178–87, on p. 87.

108. A. Buchanan, D. W. Brock, N. Daniels, D. Wilker, *From Chance to Choice* (Cambridge: Cambridge University Press, 2000), p. 11.
109. A. Palmer, 'Genetic Tests Could Prevent those Like Me Being Born at All', *Sunday Telegraph*, 12 December 2010.
110. M. Reiss and R. Straughan, *Improving Nature? The Science and Ethics of Genetic Engineering* (Cambridge: Cambridge University Press, 1996), p. 219; Y. Hashiloni-Dolev, *A Life (Un) Worthy of Living: Reproductive Genetics in Israel and Germany* (Dordrecht: Springer, 2007), p. xvi.
111. Hashiloni-Dolev, *A Life (Un) Worthy of Living: Reproductive Genetics in Israel and Germany*.

INDEX